HOLT

Earth Science

Lesson Plans

HOLT, RINEHART AND WINSTON
A Harcourt Education Company
Orlando • Austin • New York • San Diego • Toronto • London

Printed in the United States of America

ISBN 0-03-041297-8

3 4 5 6 054 08 07 06

Contents

Introduction to Earth Science 1

Earth as a System 12

Models of the Earth 22

Earth Chemistry 33

Minerals of Earth's Crust 41

Rocks 49

Resources and Energy 60

The Rock Record 73

A View of Earth's Past 84

Plate Tectonics 95

Deformation of the Crust 106

Earthquakes 114

Volcanoes 124

Weathering and Erosion 132

River Systems 145

Groundwater 155

Glaciers 162

Erosion by Wind and Waves 172

The Ocean Basins 183

Ocean Water 191

Movements of the Ocean 200

The Atmosphere 209

Water in the Atmosphere 219

Weather 230

Climate 244

Studying Space 254

Planets of the Solar System 263

Minor Bodies of the Solar System 275

The Sun 291

Stars, Galaxies, and the Universe 298

Chapter 1 Introduction to Earth Science

Chapter Opener

__ **Pre-Reading Activity, Four-Corner Fold, SE**

__ **Using the Figure, Racetrack Player, TE (GENERAL)**

Section 1 What Is Earth Science?

PACING

Regular Schedule:	**with lab(s): N/A**	**without lab(s): 2 days**
Block Schedule:	**with lab(s): N/A**	**without lab(s): 1 day**

OBJECTIVES

1. **Describe** two cultures that contributed to modern scientific study.
2. **Name** the four main branches of Earth science.
3. **Discuss** how Earth scientists help us understand the world around us.

NATIONAL SCIENCE EDUCATION STANDARDS

HNS 3c: Occasionally, there are advances in science and technology that have important and long-lasting effects on science and society. Examples of such advances include the following: Copernican revolution; Newtonian mechanics; Relativity; Geologic time scale; Plate tectonics; Atomic theory; Nuclear physics; Biological evolution; Germ theory; Industrial revolution; Molecular biology; Information and communication; Quantum theory; Galactic universe; Medical and health technology.

SAI 2a: Scientists usually inquire about how physical, living, or designed systems function. Conceptual principles and knowledge guide scientific inquiries. Historical and current scientific knowledge influence the design and interpretation of investigations and the evaluation of proposed explanations made by other scientists.

SAI 2b: Scientists conduct investigations for a wide variety of reasons. For example, they may wish to discover new aspects of the natural world, explain recently observed phenomena, or test the conclusions of prior investigations or the predictions of current theories.

KEY **SE** = Student Edition **TE** = Teacher Edition **CRF** = Chapter Resource File

SAI 2c: Scientists rely on technology to enhance the gathering and manipulation of data. New techniques and tools provide new evidence to guide inquiry and new methods to gather data, thereby contributing to the advance of science. The accuracy and precision of the data, and therefore the quality of the exploration, depends on the technology used.

SAI 2f: Results of scientific inquiry—new knowledge and methods—emerge from different types of investigations and public communication among scientists. In communicating and defending the results of scientific inquiry, arguments must be logical and demonstrate connections between natural phenomena, investigations, and the historical body of scientific knowledge. In addition, the methods and procedures that scientists used to obtain evidence must be clearly reported to enhance opportunities for further investigation.

SPSP 6a: Science and technology are essential social enterprises, but alone they can only indicate what can happen, not what should happen. The latter involves human decisions about the use of knowledge.

SPSP 6d: Individuals and society must decide on proposals involving new research and the introduction of new technologies into society. Decisions involve assessment of alternatives, risks, costs, and benefits and consideration of who benefits and who suffers, who pays and gains, and what the risks are and who bears them. Students should understand the appropriateness and value of basic questions—"What can happen?"—"What are the odds?"—and "How do scientists and engineers know what will happen?"

ST 2a: Scientists in different disciplines ask different questions, use different methods of investigation, and accept different types of evidence to support their explanations. Many scientific investigations require the contributions of individuals from different disciplines, including engineering. New disciplines of science, such as geophysics and biochemistry often emerge at the interface of two older disciplines.

ST 2b: Science often advances with the introduction of new technologies. Solving technological problems often results in new scientific knowledge. New technologies often extend the current levels of scientific understanding and introduce new areas of research.

KEY **SE** = Student Edition **TE** = Teacher Edition **CRF** = Chapter Resource File

ST 2d: Science and technology are pursued for different purposes. Scientific inquiry is driven by the desire to understand the natural world, and technological design is driven by the need to meet human needs and solve human problems. Technology, by its nature, has a more direct effect on society than science because its purpose is to solve human problems, help humans adapt, and fulfill human aspirations. Technological solutions may create new problems. Science, by its nature, answers questions that may or may not directly influence humans. Sometimes scientific advances challenge people's beliefs and practical explanations concerning various aspects of the world.

UCP 1: The natural and designed world is complex; it is too large and complicated to investigate and comprehend all at once. Scientists and students learn to define small portions for the convenience of investigation. The units of investigation can be referred to as "systems." A system is an organized group of related objects or components that form a whole. Systems can consist, for example, of organisms, machines, fundamental particles, galaxies, ideas, numbers, transportation, and education. Systems have boundaries, components, resources flow (input and output), and feedback.

FOCUS *(5 minutes)*

__ **Overview, TE** Review the objectives listed in the Student Edition. **(GENERAL)**

__ **PowerPoint® Resources** Use the customizable presentation to enhance your lecture.

__ **Bellringer, TE** Ask students why it is important to study Earth science. **(GENERAL)**

__ **Bellringer Transparency** Use this transparency as students enter the classroom and find their seats.

MOTIVATE *(10 minutes)*

__ **Discussion, Creating Calendars, TE** Students discuss why humans create calendars and how calendars enhance a culture. **(GENERAL)**

TEACH *(20 minutes)*

__ **Directed Reading, CRF** These worksheets reinforce basic concepts and vocabulary presented in the section. **(BASIC)**

__ **Using the Figure, Doing Science, TE** Students use the figure showing different scientists to explore different fields of study in Earth science. **(ADVANCED)**

__ **Reading Skill Builder, Reading Organizer, TE** Students create a reading organizer to explore one branch of Earth science. **(BASIC)**

__ **Internet Activity, Ice Cores and Climate, TE** Students research how ice core research is being used to study ancient Earth climates as a way to learn about climate change today. **(ADVANCED)**

KEY **SE** = Student Edition **TE** = Teacher Edition **CRF** = Chapter Resource File

__ **Internet Activity, Ice Cores and Climate, CRF** This worksheet directs student research for the topic presented in the Teacher Edition. **(ADVANCED)**

__ **Group Activity, Scientific Revolutions, TE** Students work in groups to research one scientific revolution. **(GENERAL)**

__ **Skill Builder, Vocabulary, TE** Students review the suffix *-ology*. **(GENERAL)**

__ **Activity, Not Your Typical Office Job, TE** Students discuss what it would be like to have a career in one of the fields of Earth science. **(GENERAL)**

__ **Interactive Tutor CD-ROM, History and Future of Earth Sciences** Students practice what they have learned using this interactive activity. **(GENERAL)**

__ **Career Focus, Field Geologist, SE** This article discusses how geologists study the earth to unlock clues to the past and to predict future events.

__ **Group Activity, Field Geology of Your Area, TE** Students research the types of rocks that underlie their area. **(GENERAL)**

CLOSE *(10 minutes)*

__ **Section Review, SE** Students answer end-of-section vocabulary, key ideas, critical thinking, and concept mapping questions. **(GENERAL)**

__ **Reteaching, Picture This, TE** Students compare pictures of scientists working in each branch of Earth science to the science as described in the chapter. **(BASIC)**

__ **Quiz, TE** Students answer 2 questions about this section. **(GENERAL)**

__ **Section Quiz, CRF** Use this quiz to assess students' understanding of the section. **(GENERAL) Also in Spanish**

__ **Spanish Section Quiz, Spanish Assessments** Use this quiz to assess students' understanding of the section. **(GENERAL)**

__ **Alternative Assessment, Role Play, TE** Students role play being scientists in one branch of Earth science. **(GENERAL)**

KEY **SE** = Student Edition **TE** = Teacher Edition **CRF** = Chapter Resource File

Lesson Plan

Section 2 Science as a Process

PACING

Regular Schedule: **with lab(s): 2 days** **without lab(s): 1 day**
Block Schedule: **with lab(s): 1 day** **without lab(s): 0.5 day**

OBJECTIVES

1. **Explain** how science is different from other forms of human endeavor.
2. **Identify** the steps that make up scientific methods.
3. **Analyze** how scientific thought changes as new information is collected.
4. **Explain** how science affects society.

NATIONAL SCIENCE EDUCATION STANDARDS

SAI 2a: Scientists usually inquire about how physical, living, or designed systems function. Conceptual principles and knowledge guide scientific inquiries. Historical and current scientific knowledge influence the design and interpretation of investigations and the evaluation of proposed explanations made by other scientists.

SAI 2b: Scientists conduct investigations for a wide variety of reasons. For example, they may wish to discover new aspects of the natural world, explain recently observed phenomena, or test the conclusions of prior investigations or the predictions of current theories.

SAI 2d: Mathematics is essential in scientific inquiry. Mathematical tools and models guide and improve the posing of questions, gathering data, constructing explanations and communicating results.

SAI 2e: Scientific explanations must adhere to criteria such as: a proposed explanation must be logically consistent; it must abide by the rules of evidence; it must be open to questions and possible modification; and it must be based on historical and current scientific knowledge.

KEY **SE** = Student Edition **TE** = Teacher Edition **CRF** = Chapter Resource File

SAI 2f: Results of scientific inquiry—new knowledge and methods—emerge from different types of investigations and public communication among scientists. In communicating and defending the results of scientific inquiry, arguments must be logical and demonstrate connections between natural phenomena, investigations, and the historical body of scientific knowledge. In addition, the methods and procedures that scientists used to obtain evidence must be clearly reported to enhance opportunities for further investigation.

SPSP 6a: Science and technology are essential social enterprises, but alone they can only indicate what can happen, not what should happen. The latter involves human decisions about the use of knowledge.

SPSP 6d: Individuals and society must decide on proposals involving new research and the introduction of new technologies into society. Decisions involve assessment of alternatives, risks, costs, and benefits and consideration of who benefits and who suffers, who pays and gains, and what the risks are and who bears them. Students should understand the appropriateness and value of basic questions—"What can happen?"—"What are the odds?"—and "How do scientists and engineers know what will happen?"

ST 2a: Scientists in different disciplines ask different questions, use different methods of investigation, and accept different types of evidence to support their explanations. Many scientific investigations require the contributions of individuals from different disciplines, including engineering. New disciplines of science, such as geophysics and biochemistry often emerge at the interface of two older disciplines.

ST 2b: Science often advances with the introduction of new technologies. Solving technological problems often results in new scientific knowledge. New technologies often extend the current levels of scientific understanding and introduce new areas of research.

ST 2d: Science and technology are pursued for different purposes. Scientific inquiry is driven by the desire to understand the natural world, and technological design is driven by the need to meet human needs and solve human problems. Technology, by its nature, has a more direct effect on society than science because its purpose is to solve human problems, help humans adapt, and fulfill human aspirations. Technological solutions may create new problems. Science, by its nature, answers questions that may or may not directly influence humans. Sometimes scientific advances challenge people's beliefs and practical explanations concerning various aspects of the world.

KEY **SE** = Student Edition **TE** = Teacher Edition **CRF** = Chapter Resource File

Lesson Plan

UCP 1: The natural and designed world is complex; it is too large and complicated to investigate and comprehend all at once. Scientists and students learn to define small portions for the convenience of investigation. The units of investigation can be referred to as "systems." A system is an organized group of related objects or components that form a whole. Systems can consist, for example, of organisms, machines, fundamental particles, galaxies, ideas, numbers, transportation, and education. Systems have boundaries, components, resources flow (input and output), and feedback.

FOCUS *(5 minutes)*

__ **Overview, TE** Review the objectives listed in the Student Edition. **(GENERAL)**

__ **PowerPoint® Resources** Use the customizable presentation to enhance your lecture.

__ **Bellringer, TE** Ask students what they think the goals of the following human endeavors are: art, architecture, philosophy, sports, and politics. **(GENERAL)**

__ **Bellringer Transparency** Use this transparency as students enter the classroom and find their seats.

MOTIVATE *(10 minutes)*

__ **Discussion, Influences on Natural Systems, TE** Have students list reasons why different plants of the same species grow at different rates. **(GENERAL)**

TEACH *(65 minutes)*

__ **Directed Reading, CRF** These worksheets reinforce basic concepts and vocabulary presented in the section. **(BASIC)**

__ **Using the Figure, Scientific Methods, TE** Students use the figure showing scientific methods to answer questions about the steps for scientific inquiry. **(GENERAL)**

__ **Discussion, Types of Reasoning, TE** Students review inductive and deductive reasoning. **(GENERAL)**

__ **Teaching Transparency 1, Sea-Floor Spreading with Worksheet** Use this Teaching Transparency and the accompanying worksheet to support your classroom discussion of sea-floor spreading.

__ **Discussion, Variables and Controls, TE** Students discuss why it is important to study only one independent variable at a time in an experiment. **(GENERAL)**

__ **Quick Lab, Making Observations, SE** Students use a candle to practice making observations. **(GENERAL)**

__ **Datasheet for Quick Lab, Making Observations, CRF** Students use the datasheet to complete the Quick Lab. **(GENERAL)**

__ **Reading Skill Builder, Reading Organizer, TE** Students create an outline listing the steps of the scientific method. **(BASIC)**

KEY SE = Student Edition TE = Teacher Edition CRF = Chapter Resource File

__ **Activity, Precision and Accuracy, TE** Students use different methods of measurement to discuss which measurements are the most and which are the least accurate and precise. **(BASIC)**

__ **Skill Builder, Math, TE** Students measure objects in the classroom and the convert the measurements to SI units. **(GENERAL)**

__ **Quick Lab, Sample Size and Accuracy, SE** Students use a deck of cards to demonstrate sample size and accuracy. **(GENERAL)**

__ **Datasheet for Quick Lab, Sample Size and Accuracy, CRF** Students use the datasheet to complete the Quick Lab. **(GENERAL)**

__ **Social Studies Connection, Margin of Error, TE** Students discuss why it is important to note the percentage of error in polls and other types of statistics. **(GENERAL)**

__ **Using the Figure, Levels of Analysis, TE** Students use the two figures of Mount Everest to discuss the differences in scale. **(BASIC)**

__ **Graphic Organizer, Chain-of-Events Chart, SE** Use the graphic organizer to help students develop their analytical skills. **(GENERAL)**

__ **Discussion, Peer Review Pressure, TE** Students discuss whether peer review is more important in science than in other fields. **(GENERAL)**

__ **Debate, Should Science Always Be Applied? TE** Students debate whether scientific discoveries should always be applied to everyday life. **(ADVANCED)**

__ **Using the Figure, Different Disciplines, TE** Students use the figures showing the extinction of the dinosaurs to answer questions about different disciplines of science. **(GENERAL)**

__ **Teaching Transparency 2, The Importance of Interdisciplinary Science with Worksheet** Use this Teaching Transparency and the accompanying worksheet to support your classroom discussion of the importance of interdisciplinary science.

__ **Inquiry Lab, Scientific Methods, SE** Students apply scientific methods as they examine a place where puddles often form after rainstorms. **(GENERAL)**

__ **Datasheet for Chapter Lab, Inquiry Lab, Scientific Methods, CRF** Students use the datasheet to complete the Chapter Lab. **(GENERAL)**

__ **Maps in Action, Geologic Features and Political Boundaries in Europe, SE** Students answer questions about a map showing the political boundaries of a part of Europe to enhance their map-reading skills. **(GENERAL)**

__ **Teaching Transparency 3, Geological Features and Political Boundaries in Europe with Worksheet** Use this Teaching Transparency and the accompanying worksheet to help students perform the Maps in Action activity.

__ **Group Activity, Making Maps** In groups, students create maps of one region of the United States. **(GENERAL)**

__ **Inquiry Lab, What's Before Your Eyes, CRF** Students observe how the rate of evaporation of a salt solution affects the formation of salt crystals. **(GENERAL)**

__ **Skills Practice Lab, Testing a Prediction, CRF** Students use water temperature to practice following procedures and to make and test predictions. **(GENERAL)**

KEY **SE** = Student Edition **TE** = Teacher Edition **CRF** = Chapter Resource File

CLOSE *(10 minutes)*

__ **Section Review, SE** Students answer end-of-section vocabulary, key ideas, critical thinking, and concept mapping questions. **(GENERAL)**

__ **Reteaching, Chart of Scientific Methods, TE** Students create a chart of the different steps in the scientific method. **(BASIC)**

__ **Quiz, TE** Students answer 2 questions about this section. **(GENERAL)**

__ **Section Quiz, CRF** Use this quiz to assess students' understanding of the section. **(GENERAL) Also in Spanish**

__ **Spanish Section Quiz, Spanish Assessments** Use this quiz to assess students' understanding of the section. **(GENERAL)**

__ **Alternative Assessment, Scientific Proposal, TE** Students write a proposal in which they describe a scientific question they would like to answer by performing an experiment. **(GENERAL)**

KEY **SE** = Student Edition **TE** = Teacher Edition **CRF** = Chapter Resource File

End of Chapter Review and Assessment

PACING

Regular Schedule:	**with lab(s): N/A**	**without lab(s): 2 days**
Block Schedule:	**with lab(s): N/A**	**without lab(s): 1 day**

___ **Chapter Review, SE** Students answer end-of-chapter vocabulary, key ideas, critical thinking, math, writing, and interpreting graphics questions. **(GENERAL)**

___ **Alternative Assessment, Frontiers of Earth Science, TE** Students think of one question they have about Earth science that could be answered by new research. **(GENERAL)**

___ **Standardized Test Prep, SE** Have students complete the standardized test prep to help them prepare for standardized tests. **(GENERAL)**

___ **Holt PuzzlePro®** Use this software to create crossword puzzles and word searches that make learning vocabulary fun. **(GENERAL)**

___ **Concept Review, CRF** Use this worksheet to assess student's understanding of the main concepts presented in the chapter. **(GENERAL)**

___ **Critical Thinking, CRF** Students use this worksheet to develop and practice critical thinking skills. **(ADVANCED)**

___ **Math Skills, CRF** This worksheet helps reinforce students' understanding of significant digits. **(GENERAL)**

___ **Graphing Skills, CRF** This worksheet helps reinforce students' understanding of line graphs. **(GENERAL)**

___ **Concept Review, Study Guide, Introduction to Earth Science** Use this worksheet to assess student's understanding of the main concepts presented in the chapter. **(GENERAL)**

___ **Concept Review, Spanish Study Guide, Introduction to Earth Science** This worksheet helps reinforce English-language learners' understanding of the main concepts presented in the chapter. **(GENERAL)**

___ **Chapter Test A, CRF** Assign questions from the appropriate test for chapter assessment. **(GENERAL)**

___ **Chapter Test B, CRF** Assign questions from the appropriate test for chapter assessment. **(ADVANCED)**

___ **Spanish Chapter Test A, Spanish Assessments** Assign questions from the appropriate test for chapter assessment. **(GENERAL)**

___ **Test Generator, One-Stop Planner** Create a customized homework assignment, quiz, or test using the HRW Test Generator program. **(GENERAL)**

___ **Test Item Listing, CRF** Use the Test Item Listing to identify questions for a customized homework, quiz, or test.

___ **PowerPoint® Resources** Use the customizable presentation to help students practice standardized test-taking skills.

KEY **SE** = Student Edition **TE** = Teacher Edition **CRF** = Chapter Resource File

OTHER RESOURCE OPTIONS

__ **HRW Earth Science Video, Segment 1, "Introduction to Earth Science"** This segment examines both agents of catastrophic change and slow, steady processes that have shaped the planet since its formation. **(GENERAL)**

__ **HRW Earth Science Video, Segment 2, "Careers in Earth Science"** This segment focuses on Earth-science professionals and explores the kinds of jobs each performs and the contributions to knowledge each makes. **(GENERAL)**

__ **NOVA Video, "Adrift on the Gulf Stream"** NOVA explores the Gulf Stream's importance to ocean life and climate from a variety of vantage points.

__ **NOVA Video, "Earthquake"** This program shows how today's advanced technology helps geologists predict earthquakes.

__ **NOVA Video, "Runaway Universe"** Join two competing teams in their quest to discover the secrets to the stars and the ultimate fate of the universe.

__ **Brain Food Video Quiz** Use this video resource to assess student's progress. **(GENERAL)**

KEY **SE** = Student Edition **TE** = Teacher Edition **CRF** = Chapter Resource File

Chapter 2 Earth as a System

Chapter Opener

__ **Pre-Reading Activity, Table Fold, SE**

__ **Using the Figure, Energy Changes, TE (GENERAL)**

Section 1 Earth: A Unique Planet

PACING

Regular Schedule:	**with lab(s): N/A**	**without lab(s): 2 days**
Block Schedule:	**with lab(s): N/A**	**without lab(s): 1 day**

OBJECTIVES

1. **Describe** the size and shape of Earth.
2. **Describe** the compositional and structural layers of Earth's interior.
3. **Identify** the possible source of Earth's magnetic field.
4. **Summarize** Newton's law of gravitation.

NATIONAL SCIENCE EDUCATION STANDARDS

PS 4b: Gravitation is a universal force that each mass exerts on any other mass. The strength of the gravitational attractive force between two masses is proportional to the masses and inversely proportional to the square of the distance between them.

UCP 1: The natural and designed world is complex; it is too large and complicated to investigate and comprehend all at once. Scientists and students learn to define small portions for the convenience of investigation. The units of investigation can be referred to as "systems." A system is an organized group of related objects or components that form a whole. Systems can consist, for example, of organisms, machines, fundamental particles, galaxies, ideas, numbers, transportation, and education. Systems have boundaries, components, resources flow (input and output), and feedback.

FOCUS *(5 minutes)*

__ **Overview, TE** Review the objectives listed in the Student Edition. **(GENERAL)**

__ **PowerPoint® Resources** Use the customizable presentation to enhance your lecture.

__ **Bellringer, TE** Students list four features of Earth. **(GENERAL)**

__ **Bellringer Transparency** Use this transparency as students enter the classroom and find their seats.

KEY **SE** = Student Edition **TE** = Teacher Edition **CRF** = Chapter Resource File

MOTIVATE *(10 minutes)*

__ **Discussion, Earth and its Moon, TE** Students list similarities and differences between Earth and its moon. **(GENERAL)**

TEACH *(20 minutes)*

__ **Directed Reading, CRF** These worksheets reinforce basic concepts and vocabulary presented in the section. **(BASIC)**

__ **Using the Figure, Earth's Interior Structure, TE** Students discuss the various layers of Earth's interior and the properties of each layer. **(BASIC)**

__ **Teaching Transparency 4, Earth's Interior with Worksheet** Use this Teaching Transparency and the accompanying worksheet to support your classroom discussion of Earth's interior structure.

__ **Physics Connection, Seismic Waves, TE** Students research longitudinal and transverse waves. **(ADVANCED)**

__ **Skill Builder, Writing, TE** Students write an essay comparing and contrasting Earth's magnetic field with that of another planet in the solar system. **(ADVANCED)**

__ **Interactive Tutor CD-ROM, Earth's Interior** Students practice what they have learned using this interactive activity. **(GENERAL)**

__ **Interactive Tutor CD-ROM, Gravity, Orbits, Motion, and Forces** Students practice what they have learned using this interactive activity. **(GENERAL)**

__ **Interactive Tutor CD-ROM, Weight, Mass, Volume, and Density** Students practice what they have learned using this interactive activity. **(GENERAL)**

CLOSE *(10 minutes)*

__ **Section Review, SE** Students answer end-of-section vocabulary, key ideas, critical thinking, and concept mapping questions. **(GENERAL)**

__ **Reteaching, Earth's Interior Structure, TE** Students identify concentric circles drawn on the board as each layer of Earth. **(BASIC)**

__ **Quiz, TE** Students answer 3 questions about this section. **(GENERAL)**

__ **Section Quiz, CRF** Use this quiz to assess students' understanding of the section. **(GENERAL) Also in Spanish**

__ **Spanish Section Quiz, Spanish Assessments** Use this quiz to assess students' understanding of the section. **(GENERAL)**

__ **Alternative Assessment, Poster Project, TE** Students create a poster comparing Earth with another planet in the solar system. **(ADVANCED)**

KEY **SE** = Student Edition **TE** = Teacher Edition **CRF** = Chapter Resource File

Section 2 Energy in the Earth System

PACING

Regular Schedule:	**with lab(s): N/A**	**without lab(s): 2 days**
Block Schedule:	**with lab(s): N/A**	**without lab(s): 1 day**

OBJECTIVES

1. **Compare** an open system with a closed system.
2. **List** the characteristics of Earth's four major spheres.
3. **Identify** the two main sources of energy in the Earth system.
4. **Identify** four processes in which matter and energy cycle on Earth.

NATIONAL SCIENCE EDUCATION STANDARDS

LS 4a: The atoms and molecules on the earth cycle among the living and nonliving components of the biosphere.

LS 5a: All matter tends toward more disorganized states. Living systems require a continuous input of energy to maintain their chemical and physical organizations. With death, and the cessation of energy input, living systems rapidly disintegrate.

LS 5b: The energy for life primarily derives from the sun. Plants capture energy by absorbing light and using it to form strong (covalent) chemical bonds between the atoms of carbon-containing (organic) molecules. These molecules can be used to assemble larger molecules with biological activity (including proteins, DNA, sugars, and fats). In addition, the energy stored in bonds between the atoms (chemical energy) can be used as sources of energy for life processes.

LS 5e: The distribution and abundance of organisms and populations in ecosystems are limited by the availability of matter and energy and the ability of the ecosystem to recycle materials.

PS 5a: The total energy of the universe is constant. Energy can be transferred by collisions in chemical and nuclear reactions, by light waves and other radiations, and in many other ways. However, it can never be destroyed. As these transfers occur, the matter involved becomes steadily less ordered.

PS 5d: Everything tends to become less organized and less orderly over time. Thus, in all energy transfers, the overall effect is that the energy is spread out uniformly. Examples are the transfer of energy from hotter to cooler objects by conduction, radiation, or convection and the warming of our surroundings when we burn fuels.

UCP 1: The natural and designed world is complex; it is too large and complicated to investigate and comprehend all at once. Scientists and students learn to define small portions for the convenience of investigation. The units of investigation can be referred to as "systems." A system is an organized group of related objects or components that form a whole. Systems can consist, for example, of organisms, machines, fundamental

KEY **SE** = Student Edition **TE** = Teacher Edition **CRF** = Chapter Resource File

particles, galaxies, ideas, numbers, transportation, and education. Systems have boundaries, components, resources flow (input and output), and feedback.

FOCUS *(5 minutes)*

__ **Overview, TE** Review the objectives listed in the Student Edition. **(GENERAL)**

__ **PowerPoint® Resources** Use the customizable presentation to enhance your lecture.

__ **Bellringer, TE** Students write down how energy and matter are added to and removed from the Earth system. **(GENERAL)**

__ **Bellringer Transparency** Use this transparency as students enter the classroom and find their seats.

MOTIVATE *(10 minutes)*

__ **Demonstration, Open and Closed Systems, TE** Students observe a pan of boiling water and identify whether it represents an open or a closed system. **(GENERAL)**

TEACH *(20 minutes)*

__ **Directed Reading, CRF** These worksheets reinforce basic concepts and vocabulary presented in the section. **(BASIC)**

__ **Reading Skill Builder, Paired Summarizing, TE** Students read the text in pairs and take turns summarizing what they have read. **(BASIC)**

__ **Teaching Transparency 5, Open and Closed Systems with Worksheet** Use this Teaching Transparency and the accompanying worksheet to support your classroom discussion of open and closed systems.

__ **Space Science Connection, Spheres of Other Planets and Moons, TE** Students research planets or large moons of planets noting which spheres they possess. **(ADVANCED)**

__ **Using the Figure, Overlapping of Spheres, TE** Students identify elements found in each of the Earth's spheres and note how they overlap. **(BASIC)**

__ **Graphic Organizer, Comparison Table, SE** Use the graphic organizer to help students develop their analytical skills. **(GENERAL)**

__ **Using the Figure, Energy Budget, TE** Students calculate how much energy is reradiated into space and how much energy is not absorbed by water and land. **(GENERAL)**

__ **Teaching Transparency 6, Earth's Energy Budget with Worksheet** Use this Teaching Transparency and the accompanying worksheet to support your classroom discussion of energy budgets.

__ **Physics Connection, The Laws of Thermodynamics, TE** Students research the history of the four laws of thermodynamics. **(ADVANCED)**

__ **Quick Lab, Effects of Solar Energy, SE** Students examine how solar energy works. **(GENERAL)**

__ **Datasheet for Quick Lab, Effects of Solar Energy, CRF** Students use the datasheet to complete the Quick Lab. **(GENERAL)**

KEY **SE** = Student Edition **TE** = Teacher Edition **CRF** = Chapter Resource File

__ **Reading Skill Builder, Reading Hint, TE** Before reading the passage, students examine the origins and meaning of the word *cycle.* **(BASIC)**

__ **Using the Figure, Nitrogen Cycle, TE** Students observe the directions of the arrows in the figure and note that nitrogen in the air is the main component of the nitrogen cycle. **(BASIC)**

__ **Teaching Transparency 7, The Nitrogen Cycle with Worksheet** Use this Teaching Transparency and the accompanying worksheet to support your classroom discussion of the nitrogen cycle.

__ **Teaching Transparency 8, The Carbon Cycle with Worksheet** Use this Teaching Transparency and the accompanying worksheet to support your classroom discussion of the carbon cycle.

__ **Group Activity, Earth's Other Cycles, TE** In groups, students research the oxygen, rock, and energy cycles of Earth's spheres. **(ADVANCED)**

__ **Inquiry Lab, Energy Transfer, CRF** Students observe how placing cold objects in water affects the transfer of thermal energy. **(GENERAL)**

__ **Making Models Lab, The Water Cycle, CRF** Students examine how water moves through the water cycle by modeling the system. **(GENERAL)**

__ **Interactive Tutor CD-ROM, Earth Systems** Students practice what they have learned using this interactive activity. **(GENERAL)**

CLOSE *(10 minutes)*

__ **Section Review, SE** Students answer end-of-section vocabulary, key ideas, critical thinking, and concept mapping questions. **(GENERAL)**

__ **Homework, Systems at Home, TE** Students list some open and closed systems found in their homes. **(GENERAL)**

__ **Reteaching, Open and Closed Systems, TE** Students identify open and closed systems from a list on the board. **(BASIC)**

__ **Quiz, TE** Students answer 3 questions about this section. **(GENERAL)**

__ **Section Quiz, CRF** Use this quiz to assess students' understanding of the section. **(GENERAL) Also in Spanish**

__ **Spanish Section Quiz, Spanish Assessments** Use this quiz to assess students' understanding of the section. **(GENERAL)**

__ **Alternative Assessment, Rates of Cycles, TE** Students research the time needed for each step to take place in the phosphorous, nitrogen, or long-term carbon cycle. **(ADVANCED)**

KEY SE = Student Edition TE = Teacher Edition CRF = Chapter Resource File

Section 3 Ecology

PACING

Regular Schedule:	**with lab(s): 2 days**	**without lab(s): 1 day**
Block Schedule:	**with lab(s): 1 day**	**without lab(s): 0.5 day**

OBJECTIVES

1. **Define** *ecosystem.*
2. **Identify** three factors that control the balance of an ecosystem.
3. **Summarize** how energy is transferred through an ecosystem.
4. **Describe** one way that ecosystems respond to environmental change.

NATIONAL SCIENCE EDUCATION STANDARDS

LS 4a: The atoms and molecules on the earth cycle among the living and nonliving components of the biosphere.

LS 4b: Energy flows through ecosystems in one direction, from photosynthetic organisms to herbivores to carnivores and decomposers.

LS 4c: Organisms both cooperate and compete in ecosystems. The interrelationships and interdependencies of these organisms may generate ecosystems that are stable for hundreds or thousands of years.

LS 4d: Living organisms have the capacity to produce populations of infinite size, but environments and resources are finite. This fundamental tension has profound effects on the interactions between organisms.

LS 4e: Human beings live within the world's ecosystems. Increasingly, humans modify ecosystems as a result of population growth, technology, and consumption. Human destruction of habitats through direct harvesting, pollution, atmospheric changes, and other factors is threatening current global stability, and if not addressed, ecosystems will be irreversibly affected.

LS 5a: All matter tends toward more disorganized states. Living systems require a continuous input of energy to maintain their chemical and physical organizations. With death, and the cessation of energy input, living systems rapidly disintegrate.

LS 5b: The energy for life primarily derives from the sun. Plants capture energy by absorbing light and using it to form strong (covalent) chemical bonds between the atoms of carbon-containing (organic) molecules. These molecules can be used to assemble larger molecules with biological activity (including proteins, DNA, sugars, and fats). In addition, the energy stored in bonds between the atoms (chemical energy) can be used as sources of energy for life processes.

LS 5c: The chemical bonds of food molecules contain energy. Energy is released when the bonds of food molecules are broken and new compounds with lower energy bonds are

KEY **SE** = Student Edition **TE** = Teacher Edition **CRF** = Chapter Resource File

formed. Cells usually store this energy temporarily in phosphate bonds of a small high-energy compound called ATP.

LS 5e: The distribution and abundance of organisms and populations in ecosystems are limited by the availability of matter and energy and the ability of the ecosystem to recycle materials.

FOCUS *(5 minutes)*

__ **Overview, TE** Review the objectives listed in the Student Edition. **(GENERAL)**

__ **PowerPoint® Resources** Use the customizable presentation to enhance your lecture.

__ **Bellringer, TE** Students write down what they think the words *ecosystem* and *ecology* mean. **(GENERAL)**

__ **Bellringer Transparency** Use this transparency as students enter the classroom and find their seats.

MOTIVATE *(10 minutes)*

__ **Using the Figure, Members of Ecosystems, TE** Students study the photograph and identify the producers. **(GENERAL)**

TEACH *(65 minutes)*

__ **Directed Reading, CRF** These worksheets reinforce basic concepts and vocabulary presented in the section. **(BASIC)**

__ **Activity, Poster Project, TE** Students create posters illustrating the dependency of predator populations on the populations of their prey. **(GENERAL)**

__ **Quick Lab, Studying Ecosystems, SE** Students study the organisms found in a small area of their choice. **(GENERAL)**

__ **Datasheet for Quick Lab, Studying Ecosystems, CRF** Students use the datasheet to complete the Quick Lab. **(GENERAL)**

__ **Using the Figure, Food Webs, TE** Students study the illustration of the food web and discuss the hierarchy shown. **(BASIC)**

__ **Teaching Transparency 9, A Food Web with Worksheet** Use this Teaching Transparency and the accompanying worksheet to support your classroom discussion of food webs.

__ **Skills Practice Lab, Testing the Conservation of Mass, SE** Students examine how matter changes state through chemical reactions. **(GENERAL)**

__ **Datasheet for Chapter Lab, Skills Practice Lab, Testing the Conservation of Mass, CRF** Students use the datasheet to complete the Chapter Lab. **(GENERAL)**

__ **Maps in Action, Concentration of Plant Life on Earth, SE** Students enhance their map reading-skills by answering questions about a map showing the concentration of plant life on Earth. **(GENERAL)**

KEY **SE** = Student Edition **TE** = Teacher Edition **CRF** = Chapter Resource File

__ **Teaching Transparency 10, Concentration of Plant Life on Earth with Worksheet** Use this Teaching Transparency and the accompanying worksheet to help students perform the Maps in Action activity.

__ **Activity, Ocean-Plant Distributions, TE** Students research types of plants found in the oceans and explain the distribution of plant life. **(GENERAL)**

__ **Impact on Society, Biological Clocks, SE** Students examine circadian rhythms, their believed origins, and the problems that arise when they are disrupted.

__ **Internet Activity, Health and Biological Clocks, TE** Students research health problems that stem from changes in human rhythms and how the problems are treated. **(ADVANCED)**

__ **Internet Activity, Health and Biological Clocks, CRF** This worksheet directs student research for the topic presented in the Teacher Edition. **(ADVANCED)**

__ **Interactive Tutor CD-ROM, Ice, Water, and Vapor** Students practice what they have learned using this interactive activity. **(GENERAL)**

__ **Interactive Tutor CD-ROM, The Water Cycle** Students practice what they have learned using this interactive activity. **(GENERAL)**

CLOSE *(10 minutes)*

__ **Section Review, SE** Students answer end-of-section vocabulary, key ideas, critical thinking, and concept mapping questions. **(GENERAL)**

__ **Reteaching, Food Chains, TE** Students arrange a list of organisms according to their place in the food chain. **(BASIC)**

__ **Quiz, TE** Students answer 2 questions about this section. **(GENERAL)**

__ **Section Quiz, CRF** Use this quiz to assess students' understanding of the section. **(GENERAL) Also in Spanish**

__ **Spanish Section Quiz, Spanish Assessments** Use this quiz to assess students' understanding of the section. **(GENERAL)**

__ **Alternative Assessment, Modeling Ecosystems, TE** Students create a model or poster showing the relationships between organisms in an ecosystem. **(ADVANCED)**

KEY **SE** = Student Edition **TE** = Teacher Edition **CRF** = Chapter Resource File

End of Chapter Review and Assessment

PACING

Regular Schedule:	**with lab(s): N/A**	**without lab(s): 2 days**
Block Schedule:	**with lab(s): N/A**	**without lab(s): 1 day**

__ **Chapter Review, SE** Students answer end-of-chapter vocabulary, key ideas, critical thinking, math, writing, and interpreting graphics questions. **(GENERAL)**

__ **Alternative Assessment, Poster Project, TE** Students create a poster showing the energy budget and food web of an ecosystem. **(GENERAL)**

__ **Standardized Test Prep, SE** Have students complete the standardized test prep to help them prepare for standardized tests. **(GENERAL)**

__ **Holt PuzzlePro®** Use this software to create crossword puzzles and word searches that make learning vocabulary fun. **(GENERAL)**

__ **Concept Review, CRF** Use this worksheet to assess student's understanding of the main concepts presented in the chapter. **(GENERAL)**

__ **Critical Thinking, CRF** Students use this worksheet to develop and practice critical thinking skills. **(ADVANCED)**

__ **Math Skills, CRF** This worksheet helps reinforce students' understanding of algebraic rearrangements. **(GENERAL)**

__ **Graphing Skills, CRF** This worksheet helps reinforce students' understanding of bar graphs. **(GENERAL)**

__ **Concept Review, Study Guide, Earth as a System** Use this worksheet to assess student's understanding of the main concepts presented in the chapter. **(GENERAL)**

__ **Concept Review, Spanish Study Guide, Earth as a System** This worksheet helps reinforce English-language learners' understanding of the main concepts presented in the chapter. **(GENERAL)**

__ **Chapter Test A, CRF** Assign questions from the appropriate test for chapter assessment. **(GENERAL)**

__ **Chapter Test B, CRF** Assign questions from the appropriate test for chapter assessment. **(ADVANCED)**

__ **Spanish Chapter Test A, Spanish Assessments** Assign questions from the appropriate test for chapter assessment. **(GENERAL)**

__ **Test Generator, One-Stop Planner** Create a customized homework assignment, quiz, or test using the HRW Test Generator program. **(GENERAL)**

__ **Test Item Listing, CRF** Use the Test Item Listing to identify questions for a customized homework, quiz, or test.

__ **PowerPoint® Resources** Use the customizable presentation to help students practice standardized test-taking skills.

KEY **SE** = Student Edition **TE** = Teacher Edition **CRF** = Chapter Resource File

OTHER RESOURCE OPTIONS

__ **Brain Food Video Quiz** Use this video resource to assess student's progress. **(GENERAL)**

KEY **SE** = Student Edition **TE** = Teacher Edition **CRF** = Chapter Resource File

Chapter 3 Models of the Earth

Chapter Opener

__ **Pre-Reading Activity, Three-Panel Flip Chart, SE**

__ **Using the Figure, Topographic Map, TE (GENERAL)**

__ **Long-Term Project, Positions of Sunrise and Sunset, SE** Students measure changes in the direction of sunrise and sunset for 8 or 9 months. **(ADVANCED)**

__ **Long-Term Project, Positions of Sunrise and Sunset, Long-Term Projects** Students use this datasheet to complete the Long-Term Project. **(ADVANCED)**

Section 1 Finding Locations on Earth

PACING

Regular Schedule:	**with lab(s): N/A**	**without lab(s): 2 days**
Block Schedule:	**with lab(s): N/A**	**without lab(s): 1 day**

OBJECTIVES

1. **Distinguish** between latitude and longitude.
2. **Explain** how latitude and longitude can be used to locate places on Earth's surface.
3. **Explain** how a magnetic compass can be used to find directions on Earth's surface.

NATIONAL SCIENCE EDUCATION STANDARDS

UCP 1: The natural and designed world is complex; it is too large and complicated to investigate and comprehend all at once. Scientists and students learn to define small portions for the convenience of investigation. The units of investigation can be referred to as "systems." A system is an organized group of related objects or components that form a whole. Systems can consist, for example, of organisms, machines, fundamental particles, galaxies, ideas, numbers, transportation, and education. Systems have boundaries, components, resources flow (input and output), and feedback.

UCP 2: Evidence consists of observations and data on which to base scientific explanations. Using evidence to understand interactions allows individuals to predict changes in natural and designed systems.
Models are tentative schemes or structures that correspond to real objects, events, or classes of events, and that have explanatory power. Models help scientists and engineers understand how things work. Models take many forms, including physical objects, plans, mental constructs, mathematical equations, and computer simulations.

SAI 2c: Scientists rely on technology to enhance the gathering and manipulation of data. New techniques and tools provide new evidence to guide inquiry and new methods to gather data, thereby contributing to the advance of science. The accuracy and precision of the data, and therefore the quality of the exploration, depends on the technology used.

KEY **SE** = Student Edition **TE** = Teacher Edition **CRF** = Chapter Resource File

SAI 2d: Mathematics is essential in scientific inquiry. Mathematical tools and models guide and improve the posing of questions, gathering data, constructing explanations and communicating results.

HNS 2a: Science distinguishes itself from other ways of knowing and from other bodies of knowledge through the use of empirical standards, logical arguments, and skepticism, as scientists strive for the best possible explanations about the natural world.

HNS 3c: Occasionally, there are advances in science and technology that have important and long-lasting effects on science and society. Examples of such advances include the following: Copernican revolution; Newtonian mechanics; Relativity; Geologic time scale; Plate tectonics; Atomic theory; Nuclear physics; Biological evolution; Germ theory; Industrial revolution; Molecular biology; Information and communication; Quantum theory; Galactic universe; Medical and health technology

HNS 3d: The historical perspective of scientific explanations demonstrates how scientific knowledge changes by evolving over time, almost always building on earlier knowledge.

FOCUS *(5 minutes)*

__ **Overview, TE** Review the objectives listed in the Student Edition. **(GENERAL)**

__ **PowerPoint® Resources** Use the customizable presentation to enhance your lecture.

__ **Bellringer, TE** Students write out descriptions of the routes they take to school. **(GENERAL)**

__ **Bellringer Transparency** Use this transparency as students enter the classroom and find their seats.

MOTIVATE *(10 minutes)*

__ **Activity, On the Grid, TE** Students use a grid on the floor of the classroom to map out various objects. **(GENERAL)**

TEACH *(20 minutes)*

__ **Directed Reading, CRF** These worksheets reinforce basic concepts and vocabulary presented in the section. **(BASIC)**

__ **Teaching Transparency 11, Parallels, Meridians, and Great Circles with Worksheet** Use this Teaching Transparency and the accompanying worksheet to support your classroom discussion of parallels, meridians, and great circles.

__ **History Connection, Early Cartographer, TE** Students research Ptolemy's contributions to Earth science. **(ADVANCED)**

__ **Skill Builder, Vocabulary, TE** Students review the origin of the word *meridian*. **(GENERAL)**

__ **Internet Activity, Magnetic Declination, TE** Students use latitude and longitude to determine the geomagnetic declination of their city. **(GENERAL)**

KEY **SE** = Student Edition **TE** = Teacher Edition **CRF** = Chapter Resource File

__ **Internet Activity, Magnetic Declination, CRF** This worksheet directs student research for the topic presented in the Teacher Edition. **(GENERAL)**

__ **Teaching Transparency 12, Magnetic Declination of the United States with Worksheet** Use this Teaching Transparency and the accompanying worksheet to support your classroom discussion of the magnetic declination of the United States.

CLOSE *(10 minutes)*

__ **Section Review, SE** Students answer end-of-section vocabulary, key ideas, critical thinking, and concept mapping questions. **(GENERAL)**

__ **Reteaching, Global Grid, TE** Students identify meridians and parallels from a ball encircled with rubber bands. **(BASIC)**

__ **Quiz, TE** Students answer 2 questions about this section. **(GENERAL)**

__ **Section Quiz, CRF** Use this quiz to assess students' understanding of the section. **(GENERAL) Also in Spanish**

__ **Spanish Section Quiz, Spanish Assessments** Use this quiz to assess students' understanding of the section. **(GENERAL)**

__ **Alternative Assessment, Cartoons Around the Globe, TE** Students create cartoons or comic strips that focus on a key concept from this section. **(GENERAL)**

KEY **SE** = Student Edition **TE** = Teacher Edition **CRF** = Chapter Resource File

Section 2 Mapping Earth's Surface

PACING

Regular Schedule:	**with lab(s): N/A**	**without lab(s): 2 days**
Block Schedule:	**with lab(s): N/A**	**without lab(s): 1 day**

OBJECTIVES

1. **Explain** two ways that scientists get data to make maps.
2. **Describe** the characteristics and uses of three types of map projections.
3. **Summarize** how to use keys, legends, and scales to read maps.

NATIONAL SCIENCE EDUCATION STANDARDS

UCP 1: The natural and designed world is complex; it is too large and complicated to investigate and comprehend all at once. Scientists and students learn to define small portions for the convenience of investigation. The units of investigation can be referred to as "systems." A system is an organized group of related objects or components that form a whole. Systems can consist, for example, of organisms, machines, fundamental particles, galaxies, ideas, numbers, transportation, and education. Systems have boundaries, components, resources flow (input and output), and feedback.

UCP 2: Evidence consists of observations and data on which to base scientific explanations. Using evidence to understand interactions allows individuals to predict changes in natural and designed systems.
Models are tentative schemes or structures that correspond to real objects, events, or classes of events, and that have explanatory power. Models help scientists and engineers understand how things work. Models take many forms, including physical objects, plans, mental constructs, mathematical equations, and computer simulations.

SAI 2c: Scientists rely on technology to enhance the gathering and manipulation of data. New techniques and tools provide new evidence to guide inquiry and new methods to gather data, thereby contributing to the advance of science. The accuracy and precision of the data, and therefore the quality of the exploration, depends on the technology used.

SAI 2d: Mathematics is essential in scientific inquiry. Mathematical tools and models guide and improve the posing of questions, gathering data, constructing explanations and communicating results.

HNS 2a: Science distinguishes itself from other ways of knowing and from other bodies of knowledge through the use of empirical standards, logical arguments, and skepticism, as scientists strive for the best possible explanations about the natural world.

HNS 3c: Occasionally, there are advances in science and technology that have important and long-lasting effects on science and society. Examples of such advances include the following: Copernican revolution; Newtonian mechanics; Relativity; Geologic time scale; Plate tectonics; Atomic theory; Nuclear physics; Biological evolution; Germ theory;

KEY **SE** = Student Edition **TE** = Teacher Edition **CRF** = Chapter Resource File

Lesson Plan

Industrial revolution; Molecular biology; Information and communication; Quantum theory; Galactic universe; Medical and health technology

FOCUS *(5 minutes)*

__ **Overview, TE** Review the objectives listed in the Student Edition. **(GENERAL)**

__ **PowerPoint® Resources** Use the customizable presentation to enhance your lecture.

__ **Bellringer, TE** Students write paragraphs summarizing different kinds of maps that they've seen. **(GENERAL)**

__ **Bellringer Transparency** Use this transparency as students enter the classroom and find their seats.

MOTIVATE *(10 minutes)*

__ **Group Activity, School Map, TE** In groups, students make maps of an area of their school. **(GENERAL)**

TEACH *(20 minutes)*

__ **Directed Reading, CRF** These worksheets reinforce basic concepts and vocabulary presented in the section. **(BASIC)**

__ **Quick Lab, Making Projections, SE** Students use bowls and flashlights to illustrate projection distortion. **(GENERAL)**

__ **Datasheet for Quick Lab, Making Projections, CRF** Students use the datasheet to complete the Quick Lab. **(GENERAL)**

__ **Activity, Map A-Peel, TE** Students draw maps on the skins of oranges or grapefruits to understand map distortion. **(GENERAL)**

__ **Discussion, Round Earth, Flat Maps, TE** Students discuss why a globe is the only completely accurate map. **(GENERAL)**

__ **Teaching Transparency 13, Types of Map Projections with Worksheet** Use this Teaching Transparency and the accompanying worksheet to support your classroom discussion of map projections.

__ **Group Activity, Maps and Globes, TE** In groups, students compare the sizes of land masses on different maps. **(GENERAL)**

__ **Reading Skill Builder, Reading Organizer, TE** Students are encouraged to take notes while reading this section. **(BASIC)**

__ **Geography Connection, Mercator Projections, TE** Students research the cartographer, Gerhard Kramer. **(GENERAL)**

__ **Demonstration, Sun and Shadow, TE** Use a stick and sunlight to demonstrate the origins of the cardinal directions. **(GENERAL)**

__ **Group Activity, Chart Your Course, TE** Students use survey maps and compasses to chart a course and locate their destination. **(ADVANCED)**

KEY SE = Student Edition TE = Teacher Edition CRF = Chapter Resource File

__ **Using the Figure, Planning a Trip, TE** Students use the map legend to plan a route to a nearby town shown on the map. **(GENERAL)**

__ **Eye on the Environment, Mapping Life on Earth, SE** Biodiversity maps and their helpfulness in protecting both known and unknown species are discussed.

__ **Group Activity, Biodiversity Maps, TE** Students make biodiversity maps after plotting a small area and cataloging the plant species found there. **(GENERAL)**

__ **Inquiry Lab, Scale the School, CRF** Students create small-scale maps of the area around their school. **(GENERAL)**

__ **Making Models Lab, Remote Sensing, CRF** Students make maps of a hidden landscape to see how remote-sensing technology works. **(GENERAL)**

__ **Interactive Tutor CD-ROM, Mapping and Technology** Students practice what they have learned using this interactive activity. **(GENERAL)**

__ **Interactive Tutor CD-ROM, Satellite Measurements** Students practice what they have learned using this interactive activity. **(GENERAL)**

CLOSE *(10 minutes)*

__ **Section Review, SE** Students answer end-of-section vocabulary, key ideas, critical thinking, and concept mapping questions. **(GENERAL)**

__ **Reteaching, Using the Legend, TE** Students compare different types of maps of the same area. **(BASIC)**

__ **Quiz, TE** Students answer 3 questions about this section. **(GENERAL)**

__ **Section Quiz, CRF** Use this quiz to assess students' understanding of the section. **(GENERAL) Also in Spanish**

__ **Spanish Section Quiz, Spanish Assessments** Use this quiz to assess students' understanding of the section. **(GENERAL)**

__ **Alternative Assessment, Treasure Hunt, TE** In groups, student draw maps of the school showing where treasures that they have hidden are located. **(ADVANCED)**

KEY **SE** = Student Edition **TE** = Teacher Edition **CRF** = Chapter Resource File

Section 3 Types of Maps

PACING

Regular Schedule: **with lab(s): 2 days** **without lab(s): 1 day**
Block Schedule: **with lab(s): 1 day** **without lab(s): 0.5 day**

OBJECTIVES

1. **Explain** how elevation and topography are shown on a map.
2. **Describe** three types of information shown in geologic maps.
3. **Identify** two uses of soil maps.

NATIONAL SCIENCE EDUCATION STANDARDS

UCP 1: The natural and designed world is complex; it is too large and complicated to investigate and comprehend all at once. Scientists and students learn to define small portions for the convenience of investigation. The units of investigation can be referred to as "systems." A system is an organized group of related objects or components that form a whole. Systems can consist, for example, of organisms, machines, fundamental particles, galaxies, ideas, numbers, transportation, and education. Systems have boundaries, components, resources flow (input and output), and feedback.

UCP 2: Evidence consists of observations and data on which to base scientific explanations. Using evidence to understand interactions allows individuals to predict changes in natural and designed systems.
Models are tentative schemes or structures that correspond to real objects, events, or classes of events, and that have explanatory power. Models help scientists and engineers understand how things work. Models take many forms, including physical objects, plans, mental constructs, mathematical equations, and computer simulations.

SAI 2c: Scientists rely on technology to enhance the gathering and manipulation of data. New techniques and tools provide new evidence to guide inquiry and new methods to gather data, thereby contributing to the advance of science. The accuracy and precision of the data, and therefore the quality of the exploration, depends on the technology used.

SAI 2d: Mathematics is essential in scientific inquiry. Mathematical tools and models guide and improve the posing of questions, gathering data, constructing explanations and communicating results.

HNS 2a: Science distinguishes itself from other ways of knowing and from other bodies of knowledge through the use of empirical standards, logical arguments, and skepticism, as scientists strive for the best possible explanations about the natural world.

HNS 3c: Occasionally, there are advances in science and technology that have important and long-lasting effects on science and society. Examples of such advances include the following: Copernican revolution; Newtonian mechanics; Relativity; Geologic time scale; Plate tectonics; Atomic theory; Nuclear physics; Biological evolution; Germ theory;

KEY **SE** = Student Edition **TE** = Teacher Edition **CRF** = Chapter Resource File

Industrial revolution; Molecular biology; Information and communication; Quantum theory; Galactic universe; Medical and health technology

FOCUS *(5 minutes)*

__ **Overview, TE** Review the objectives listed in the Student Edition. **(GENERAL)**

__ **PowerPoint® Resources** Use the customizable presentation to enhance your lecture.

__ **Bellringer, TE** Students draw pictures of a hill, mountain, or valley in profile to show differences in elevation. **(GENERAL)**

__ **Bellringer Transparency** Use this transparency as students enter the classroom and find their seats.

MOTIVATE *(10 minutes)*

__ **Group Activity, Terrain Models, TE** Students make relief models of a landform of their choice. **(GENERAL)**

TEACH *(65 minutes)*

__ **Directed Reading, CRF** These worksheets reinforce basic concepts and vocabulary presented in the section. **(BASIC)**

__ **Teaching Transparency 14, Topographic Maps with Worksheet** Use this Teaching Transparency and the accompanying worksheet to support your classroom discussion of topographic maps.

__ **Quick Lab, Topographic Maps, SE** Students make a mountain out of clay and convert the model to a topographic map. **(GENERAL)**

__ **Datasheet for Quick Lab, Topographic Maps, CRF** Students use the datasheet to complete the Quick Lab. **(GENERAL)**

__ **Skill Builder, Math, TE** Students calculate the elevations that contour lines represent on maps. **(GENERAL)**

__ **Geography Connection, Quadrangles, TE** Students use the Internet to find a map of their local area. **(GENERAL)**

__ **Reading Skill Builder, Reading Hint, TE** Students discuss the origins of the word *topography*. **(BASIC)**

__ **Discussion, Comprehension Check, TE** Students summarize how color is used on topographic maps. **(BASIC)**

__ **History Connection, Public Lands, TE** Students compare topographic maps of eastern states that are not affected by The Land Ordinance System of 1785. **(ADVANCED)**

__ **Using the Figure, Map Layers, TE** Students identify the various symbols used on geologic maps. **(GENERAL)**

__ **Graphic Organizer, Spider Map, SE** Use the graphic organizer to help students develop their analytical skills. **(GENERAL)**

KEY SE = Student Edition TE = Teacher Edition CRF = Chapter Resource File

__ **Discussion, Value of Geologic Maps, TE** Students speculate how developers and engineers might use geologic maps. **(GENERAL)**

__ **Group Activity, Local Soil Maps, TE** Students collect soil samples and use them to map locations of soil types. **(ADVANCED)**

__ **Making Models Lab, Contour Maps: Island Construction, SE** Students make contour maps of an island out of clay. **(GENERAL)**

__ **Datasheet for Chapter Lab, Making Models Lab, Contour Maps: Island Construction, CRF** Students use the datasheet to complete the Chapter Lab. **(GENERAL)**

__ **Mapping Expeditions, Journey to Red River, SE** Students use a real USGS topographic map to practice their map-reading skills. **(GENERAL)**

__ **Maps in Action, Topographic Map of the Desolation Watershed, SE** Students enhance their map-reading skills by answering questions about a graphic map showing the Desolation watershed. **(GENERAL)**

__ **Teaching Transparency 15, Topographic Map of the Desolation Watershed with Worksheet** Use this Teaching Transparency and the accompanying worksheet to help students perform the Maps in Action activity.

__ **Activity, Local Watershed, TE** Students use topographic maps to determine watershed boundaries for their area. **(GENERAL)**

CLOSE *(10 minutes)*

__ **Section Review, SE** Students answer end-of-section vocabulary, key ideas, critical thinking, and concept mapping questions. **(GENERAL)**

__ **Reteaching, Mapping Vocabulary, TE** Students write definitions for the key terms in this section. **(BASIC)**

__ **Quiz, TE** Students answer 4 questions about this section. **(GENERAL)**

__ **Section Quiz, CRF** Use this quiz to assess students' understanding of the section. **(GENERAL) Also in Spanish**

__ **Spanish Section Quiz, Spanish Assessments** Use this quiz to assess students' understanding of the section. **(GENERAL)**

__ **Alternative Assessment, Song Lyrics, TE** Students write songs about how different types of maps convey information. **(GENERAL)**

KEY **SE** = Student Edition **TE** = Teacher Edition **CRF** = Chapter Resource File

End of Chapter Review and Assessment

PACING

Regular Schedule: **with lab(s): N/A** **without lab(s): 2 days**
Block Schedule: **with lab(s): N/A** **without lab(s): 1 day**

__ **Chapter Review, SE** Students answer end-of-chapter vocabulary, key ideas, critical thinking, math, writing, and interpreting graphics questions. **(GENERAL)**

__ **Alternative Assessment, Mapmaker, Mapmaker, TE** In pairs, students make maps showing how to reach a destination in their neighborhood. **(ADVANCED)**

__ **Standardized Test Prep, SE** Have students complete the standardized test prep to help them prepare for standardized tests. **(GENERAL)**

__ **Holt PuzzlePro®** Use this software to create crossword puzzles and word searches that make learning vocabulary fun. **(GENERAL)**

__ **Concept Review, CRF** Use this worksheet to assess student's understanding of the main concepts presented in the chapter. **(GENERAL)**

__ **Critical Thinking, CRF** Students use this worksheet to develop and practice critical thinking skills. **(ADVANCED)**

__ **Math Skills, CRF** This worksheet helps reinforce students' understanding of exponents. **(GENERAL)**

__ **Graphing Skills, CRF** This worksheet helps reinforce students' understanding of line graphs. **(GENERAL)**

__ **Concept Review, Study Guide, Models of the Earth** Use this worksheet to assess student's understanding of the main concepts presented in the chapter. **(GENERAL)**

__ **Concept Review, Spanish Study Guide, Models of the Earth** This worksheet helps reinforce English-language learners' understanding of the main concepts presented in the chapter. **(GENERAL)**

__ **Chapter Test A, CRF** Assign questions from the appropriate test for chapter assessment. **(GENERAL)**

__ **Chapter Test B, CRF** Assign questions from the appropriate test for chapter assessment. **(ADVANCED)**

__ **Spanish Chapter Test A, Spanish Assessments** Assign questions from the appropriate test for chapter assessment. **(GENERAL)**

__ **Test Generator, One-Stop Planner** Create a customized homework assignment, quiz, or test using the HRW Test Generator program. **(GENERAL)**

__ **Test Item Listing, CRF** Use the Test Item Listing to identify questions for a customized homework, quiz, or test.

__ **PowerPoint® Resources** Use the customizable presentation to help students practice standardized test-taking skills.

KEY **SE** = Student Edition **TE** = Teacher Edition **CRF** = Chapter Resource File

OTHER RESOURCE OPTIONS

__ **Brain Food Video Quiz** Use this video resource to assess student's progress. (**GENERAL**)

KEY **SE** = Student Edition **TE** = Teacher Edition **CRF** = Chapter Resource File

Lesson Plan

Chapter 4 Earth Chemistry

Chapter Opener

__ **Pre-Reading Activity, Double Door, SE**

__ **Using the Figure, Champagne Pool, TE (GENERAL)**

Section 1 Matter

PACING

Regular Schedule:	**with lab(s): 2 days**	**without lab(s): 1 day**
Block Schedule:	**with lab(s): 1 day**	**without lab(s): 0.5 day**

OBJECTIVES

1. **Compare** chemical properties and physical properties of matter.
2. **Describe** the basic structure of an atom.
3. **Compare** atomic number, mass number, and atomic mass.
4. **Define** *isotope*.
5. **Describe** the arrangement of elements in the periodic table.

NATIONAL SCIENCE EDUCATION STANDARDS

HNS 3c: Occasionally, there are advances in science and technology that have important and long-lasting effects on science and society. Examples of such advances include the following: Copernican revolution; Newtonian mechanics; Relativity; Geologic time scale; Plate tectonics; Atomic theory; Nuclear physics; Biological evolution; Germ theory; Industrial revolution; Molecular biology; Information and communication; Quantum theory; Galactic universe; Medical and health technology.

PS 1a: Matter is made of minute particles called atoms, and atoms are composed of even smaller components. These components have measurable properties, such as mass and electrical charge. Each atom has a positively charged nucleus surrounded by negatively charged electrons. The electric force between the nucleus and electrons holds the atom together.

PS 1b: The atom's nucleus is composed of protons and neutrons, which are much more massive than electrons. When an element has atoms that differ in the number of neutrons, these atoms are called different isotopes of the element.

KEY **SE** = Student Edition **TE** = Teacher Edition **CRF** = Chapter Resource File

Lesson Plan

PS 2b: An element is composed of a single type of atom. When elements are listed in order according to the number of protons (called the atomic number), repeating patterns of physical and chemical properties identify families of elements with similar properties. This "Periodic Table" is a consequence of the repeating pattern of outermost electrons and their permitted energies.

FOCUS *(5 minutes)*

__ **Overview, TE** Review the objectives listed in the Student Edition. **(GENERAL)**

__ **PowerPoint® Resources** Use the customizable presentation to enhance your lecture.

__ **Bellringer, TE** Ask students to give examples of matter around them and to list some characteristics of matter. **(GENERAL)**

__ **Bellringer Transparency** Use this transparency as students enter the classroom and find their seats.

MOTIVATE *(10 minutes)*

__ **Using the Figure, Earth's Elements, TE** Students use the pie chart figure to identify the three most common elements in Earth's continental crust and their respective percentages. **(GENERAL)**

TEACH *(65 minutes)*

__ **Directed Reading, CRF** These worksheets reinforce basic concepts and vocabulary presented in the section. **(BASIC)**

__ **Graphic Organizer, Comparison Table, SE** Use the graphic organizer to help students develop their analytical skills. **(GENERAL)**

__ **Reading Skill Builder, Paired Summarizing, TE** Students read the text in pairs and take turns summarizing what they have read. **(BASIC)**

__ **Quick Lab, Using the Periodic Table, SE** Students use the periodic table to collect and classify data. **(GENERAL)**

__ **Datasheet for Quick Lab, Using the Periodic Table, CRF** Students use the datasheet to complete the Quick Lab. **(GENERAL)**

__ **Activity, Poster Project, TE** Have each student create a poster for an element from the periodic table. **(BASIC)**

__ **Teaching Transparency 16, Atomic Number and Atomic Mass with Worksheet** Use this Teaching Transparency and the accompanying worksheet to support your classroom discussion of atomic number and atomic mass.

__ **Astronomy Connection, Forming Elements, TE** Students research elements in an exploding Type II supernova and compare the results with the same elements on Earth. **(ADVANCED)**

KEY **SE** = Student Edition **TE** = Teacher Edition **CRF** = Chapter Resource File

__ **Teaching Transparency 17, The Periodic Table with Worksheet** Use this Teaching Transparency and the accompanying worksheet to support your classroom discussion of the periodic table.

__ **Using the Figure, Radioactive Elements, TE** Students identify the lightest radioactive element by name and atomic number. **(GENERAL)**

__ **Inquiry Lab, Physical Properties of Elements, SE** Students identify samples of various metals by collecting data. **(GENERAL)**

__ **Datasheet for Chapter Lab, Inquiry Lab, Physical Properties of Elements, CRF** Students use the datasheet to complete the Chapter Lab. **(GENERAL)**

__ **Maps in Action, Element Resources in the United States, SE** Students enhance their map-reading skills by answering questions about a map showing the distribution of elements that are used as resources in the United States. **(GENERAL)**

__ **Teaching Transparency 20, Element Resources in the United States with Worksheet** Use this Teaching Transparency and the accompanying worksheet to help students perform the Maps in Action activity.

__ **Mapping Expeditions, Buried Treasure, SE** Students use maps to learn about where resources are located in the U.S. and how this distribution affects population. **(GENERAL)**

__ **Skills Practice Lab, Flame Tests, CRF** Students use the flame test to identify unknown compounds. **(GENERAL)**

__ **Interactive Tutor CD-ROM, What's the Matter? (Atomic Theory and Nuclear Energy)** Students practice what they have learned using this interactive activity. **(GENERAL)**

CLOSE *(10 minutes)*

__ **Section Review, SE** Students answer end-of-section vocabulary, key ideas, critical thinking, and concept mapping questions. **(GENERAL)**

__ **Homework, Naming the Elements, TE** Each student researches the origin of five elements from the periodic table. **(ADVANCED)**

__ **Reteaching, Atomic and Mass Numbers, TE** Students use the periodic table to identify elements. **(BASIC)**

__ **Quiz, TE** Students answer 3 questions about this section. **(GENERAL)**

__ **Section Quiz, CRF** Use this quiz to assess students' understanding of the section. **(GENERAL) Also in Spanish**

__ **Spanish Section Quiz, Spanish Assessments** Use this quiz to assess students' understanding of the section. **(GENERAL)**

__ **Alternative Assessment, Quiz Show, TE** Students create a game show in which they have to identify elements. **(ADVANCED)**

KEY **SE** = Student Edition **TE** = Teacher Edition **CRF** = Chapter Resource File

Section 2 Combinations of Atoms

PACING

Regular Schedule:	**with lab(s): N/A**	**without lab(s): 2 days**
Block Schedule:	**with lab(s): N/A**	**without lab(s): 1 day**

OBJECTIVES

1. **Define** *compound* and *molecule.*
2. **Interpret** chemical formulas.
3. **Describe** two ways that electrons form chemical bonds between atoms.
4. **Explain** the differences between compounds and mixtures.

NATIONAL SCIENCE EDUCATION STANDARDS

PS 2a: Atoms interact with one another by transferring or sharing electrons that are furthest from the nucleus. These outer electrons govern the chemical properties of the element.

PS 2e: Solids, liquids, and gases differ in the distances and angles between molecules or atoms and therefore the energy that binds them together. In solids the structure is nearly rigid; in liquids molecules or atoms move around each other but do not move apart; and in gases molecules or atoms move almost independently of each other and are mostly far apart.

FOCUS *(5 minutes)*

__ **Overview, TE** Review the objectives listed in the Student Edition. **(GENERAL)**

__ **PowerPoint® Resources** Use the customizable presentation to enhance your lecture.

__ **Bellringer, TE** Have students list three familiar compounds that each consist of two elements. **(GENERAL)**

__ **Bellringer Transparency** Use this transparency as students enter the classroom and find their seats.

MOTIVATE *(10 minutes)*

__ **Activity, Composition of Compounds, TE** Students use the periodic table to identify elements. **(GENERAL)**

KEY **SE** = Student Edition **TE** = Teacher Edition **CRF** = Chapter Resource File

Lesson Plan

TEACH *(20 minutes)*

__ **Directed Reading, CRF** These worksheets reinforce basic concepts and vocabulary presented in the section. **(BASIC)**

__ **Using the Figure, Describing Reactions, TE** Students study the figure showing a chemical equation to determine if the equation is balanced. **(GENERAL)**

__ **Reading Skill Builder, Reading Hint, TE** Students balance a chemical equation while reading the section. **(BASIC)**

__ **Teaching Transparency 18, Balancing Equations with Worksheet** Use this Teaching Transparency and the accompanying worksheet to support your classroom discussion of balancing equations.

__ **Physics Connection, Energy Needs, TE** Students look up the amounts of energy required for melting and boiling water and lead. **(GENERAL)**

__ **Skill Builder, Math, TE** Students identify the most stable ions in a list of elements. **(GENERAL)**

__ **Quick Lab, Compounds, SE** Students add vinegar to two compounds and record their observations. **(GENERAL)**

__ **Datasheet for Quick Lab, Compounds, CRF** Students use the datasheet to complete the Quick Lab. **(GENERAL)**

__ **Teaching Transparency 19, Ionic Bonds and Covalent Bonds with Worksheet** Use this Teaching Transparency and the accompanying worksheet to support your classroom discussion of ionic bonds and covalent bonds.

__ **Science and Technology, The Smallest Particles, SE** The study of subatomic particles and its application to Earth science is discussed.

__ **Internet Activity, High-Energy Physics, TE** Students research the field of high-energy physics. **(ADVANCED)**

__ **Internet Activity, High-Energy Physics, CRF** This worksheet directs student research for the topic presented in the Teacher Edition. **(ADVANCED)**

__ **Inquiry Lab, Ionic and Covalent Conductivity, CRF** Students determine whether compounds are ionic or covalent based on how they conduct electricity. **(GENERAL)**

__ **Interactive Tutor CD-ROM, Matter and Minerals** Students practice what they have learned using this interactive activity. **(GENERAL)**

CLOSE *(10 minutes)*

__ **Section Review, SE** Students answer end-of-section vocabulary, key ideas, critical thinking, and concept mapping questions. **(GENERAL)**

__ **Homework, Atomic and Ionic Radius, TE** Students research atomic and ionic radii for a single row of the periodic table and present their findings. **(ADVANCED)**

__ **Reteaching, Types of Compounds, TE** Students identify compound formulas as either ionic or covalent. **(BASIC)**

KEY **SE** = Student Edition **TE** = Teacher Edition **CRF** = Chapter Resource File

Lesson Plan

__ **Quiz, TE** Students answer 3 questions about this section. **(GENERAL)**

__ **Section Quiz, CRF** Use this quiz to assess students' understanding of the section. **(GENERAL) Also in Spanish**

__ **Spanish Section Quiz, Spanish Assessments** Use this quiz to assess students' understanding of the section. **(GENERAL)**

__ **Alternative Assessment, Balancing Chemical Equations, TE** Students show the steps needed to balance chemical equations. **(ADVANCED)**

KEY **SE** = Student Edition **TE** = Teacher Edition **CRF** = Chapter Resource File

End of Chapter Review and Assessment

PACING

Regular Schedule:	**with lab(s): N/A**	**without lab(s): 2 days**
Block Schedule:	**with lab(s): N/A**	**without lab(s): 1 day**

__ **Chapter Review, SE** Students answer end-of-chapter vocabulary, key ideas, critical thinking, math, writing, and interpreting graphics questions. **(GENERAL)**

__ **Alternative Assessment, Poster Project, TE** Students create posters showing the processes by which atoms bond together ionically and covalently. **(GENERAL)**

__ **Standardized Test Prep, SE** Have students complete the standardized test prep to help them prepare for standardized tests. **(GENERAL)**

__ **Holt PuzzlePro®** Use this software to create crossword puzzles and word searches that make learning vocabulary fun. **(GENERAL)**

__ **Concept Review, CRF** Use this worksheet to assess student's understanding of the main concepts presented in the chapter. **(GENERAL)**

__ **Critical Thinking, CRF** Students use this worksheet to develop and practice critical thinking skills. **(ADVANCED)**

__ **Math Skills, CRF** This worksheet helps reinforce students' understanding of scientific notation. **(GENERAL)**

__ **Graphing Skills, CRF** This worksheet helps reinforce students' understanding of ternary diagrams. **(GENERAL)**

__ **Concept Review, Study Guide, Earth Chemistry** Use this worksheet to assess student's understanding of the main concepts presented in the chapter. **(GENERAL)**

__ **Concept Review, Spanish Study Guide, Earth Chemistry** This worksheet helps reinforce English-language learners' understanding of the main concepts presented in the chapter. **(GENERAL)**

__ **Chapter Test A, CRF** Assign questions from the appropriate test for chapter assessment. **(GENERAL)**

__ **Chapter Test B, CRF** Assign questions from the appropriate test for chapter assessment. **(ADVANCED)**

__ **Spanish Chapter Test A, Spanish Assessments** Assign questions from the appropriate test for chapter assessment. **(GENERAL)**

__ **Test Generator, One-Stop Planner** Create a customized homework assignment, quiz, or test using the HRW Test Generator program. **(GENERAL)**

__ **Test Item Listing, CRF** Use the Test Item Listing to identify questions for a customized homework, quiz, or test.

__ **PowerPoint® Resources** Use the customizable presentation to help students practice standardized test-taking skills.

KEY **SE** = Student Edition **TE** = Teacher Edition **CRF** = Chapter Resource File

OTHER RESOURCE OPTIONS

___ **HRW Earth Science Video, Segment 7, "Chemical Elements"** The objective of this segment is to provide a simplified overview of the elemental composition of matter. **(GENERAL)**

___ **Brain Food Video Quiz** Use this video resource to assess student's progress. **(GENERAL)**

KEY **SE** = Student Edition **TE** = Teacher Edition **CRF** = Chapter Resource File

Chapter 5 Minerals of Earth's Crust

Chapter Opener

__ **Pre-Reading Activity, Double Door, SE**

__ **Using the Figure, Quartz Quest, TE (GENERAL)**

Section 1 What Is a Mineral?

PACING

Regular Schedule:	**with lab(s): N/A**	**without lab(s): 2 days**
Block Schedule:	**with lab(s): N/A**	**without lab(s): 1 day**

OBJECTIVES

1. **Define** *mineral.*
2. **Compare** the two main groups of minerals.
3. **Identify** the six types of silicate crystalline structures.
4. **Describe** three common nonsilicate crystalline structures.

NATIONAL SCIENCE EDUCATION STANDARDS

PS 2e: Solids, liquids, and gases differ in the distances and angles between molecules or atoms and therefore the energy that binds them together. In solids the structure is nearly rigid; in liquids molecules or atoms move around each other but do not move apart; and in gases molecules or atoms move almost independently of each other and are mostly far apart.

PS 2f: Carbon atoms can bond to one another in chains, rings, and branching networks to form a variety of structures, including synthetic polymers, oils, and the large molecules essential to life.

FOCUS *(5 minutes)*

__ **Overview, TE** Review the objectives listed in the Student Edition. **(GENERAL)**

__ **PowerPoint® Resources** Use the customizable presentation to enhance your lecture.

__ **Bellringer, TE** Ask students to define what a mineral is and give an example. **(GENERAL)**

__ **Bellringer Transparency** Use this transparency as students enter the classroom and find their seats.

KEY **SE** = Student Edition **TE** = Teacher Edition **CRF** = Chapter Resource File

MOTIVATE *(10 minutes)*

__ **Group Activity, Mineral—Yes or No? TE** Students identify whether or not various samples are minerals. **(GENERAL)**

TEACH *(20 minutes)*

__ **Directed Reading, CRF** These worksheets reinforce basic concepts and vocabulary presented in the section. **(BASIC)**

__ **Teaching Transparency 21, Characteristics of Minerals with Worksheet** Use this Teaching Transparency and the accompanying worksheet to support your classroom discussion of characteristics of minerals.

__ **Internet Activity, Mineral Formation, TE** Students research the three basic ways that minerals form. **(ADVANCED)**

__ **Internet Activity, Mineral Formation, CRF** This worksheet directs student research for the topic presented in the Teacher Edition. **(ADVANCED)**

__ **Activity, Acid Test, TE** Students test samples to identify a carbonate and a noncarbonate. **(BASIC)**

__ **Skill Builder, Vocabulary, TE** Students review the word *crystal.* **(GENERAL)**

__ **Quick Lab, Modeling Tetrahedra, SE** Students use toothpicks and marshmallows to build tetrahedra chains. **(GENERAL)**

__ **Datasheet for Quick Lab, Modeling Tetrahedra, CRF** Students use the datasheet to complete the Quick Lab. **(GENERAL)**

__ **Group Activity, Growing Crystals, TE** In pairs, students grow crystals of sodium carbonate or sodium chloride. **(BASIC)**

__ **Reading Skill Builder, Paired Summarizing, TE** Students read the text in pairs and take turns summarizing what they have read. **(BASIC)**

__ **Teaching Transparency 22, Structures of Silicate Minerals with Worksheet** Use this Teaching Transparency and the accompanying worksheet to support your classroom discussion of the structures of silicate minerals.

__ **Maps in Action, Rock and Mineral Production in the United States, SE** Students enhance their map reading-skills by answering questions about a map showing the distribution of rock and mineral production in the United States. **(GENERAL)**

__ **Teaching Transparency 25, Rock and Mineral Production in the United States with Worksheet** Use this Teaching Transparency and the accompanying worksheet to help students perform the Maps in Action activity.

__ **Activity, Finding Mineral Deposits, TE** Students research the methods used to locate and assess mineral deposits. **(GENERAL)**

__ **Internet Activity, Mining Impacts, TE** Students investigate the past and present environmental effects of mining in their state. **(GENERAL)**

__ **Internet Activity, Mining Effects, CRF** This worksheet directs student research for the topic presented in the Teacher Edition. **(GENERAL)**

KEY **SE** = Student Edition **TE** = Teacher Edition **CRF** = Chapter Resource File

__ **Career Focus, Mining Engineer, SE** A mining engineer describes the purpose and highlights of her job and how it applies to the various kinds of mines and resources.

__ **Skills Practice Lab, Copper Recovery, CRF** Students use cementation to recover solid copper from a copper sulfate solution. **(GENERAL)**

CLOSE *(10 minutes)*

__ **Section Review, SE** Students answer end-of-section vocabulary, key ideas, critical thinking, and concept mapping questions. **(GENERAL)**

__ **Homework, Mineral Names, TE** Students research the origins of mineral names. **(ADVANCED)**

__ **Reteaching, Name That Mineral, TE** Students quiz each other about the differences between silicate and nonsilicate minerals. **(BASIC)**

__ **Quiz, TE** Students answer 3 questions about this section. **(GENERAL)**

__ **Section Quiz, CRF** Use this quiz to assess students' understanding of the section. **(GENERAL) Also in Spanish**

__ **Spanish Section Quiz, Spanish Assessments** Use this quiz to assess students' understanding of the section. **(GENERAL)**

__ **Alternative Assessment, Flowchart, TE** Students create a flowchart that classifies and describes the two main groups of minerals and their major subgroups. **(GENERAL)**

KEY **SE** = Student Edition **TE** = Teacher Edition **CRF** = Chapter Resource File

Section 2 Identifying Minerals

PACING

Regular Schedule:	**with lab(s): 2 days**	**without lab(s): 1 day**
Block Schedule:	**with lab(s): 1 day**	**without lab(s): 0.5 day**

OBJECTIVES

1. **Describe** seven physical properties that help distinguish one mineral from another.
2. **List** five special properties that may help identify certain minerals.

NATIONAL SCIENCE EDUCATION STANDARDS

PS 2c: Bonds between atoms are created when electrons are paired up by being transferred or shared. A substance composed of a single kind of atom is called an element. The atoms may be bonded together into molecules or crystalline solids. A compound is formed when two or more kinds of atoms bind together chemically.

PS 2f: Carbon atoms can bond to one another in chains, rings, and branching networks to form a variety of structures, including synthetic polymers, oils, and the large molecules essential to life.

FOCUS *(5 minutes)*

___ **Overview, TE** Review the objectives listed in the Student Edition. **(GENERAL)**

___ **PowerPoint® Resources** Use the customizable presentation to enhance your lecture.

___ **Bellringer, TE** Ask students to list properties they might use to tell one mineral from another. **(GENERAL)**

___ **Bellringer Transparency** Use this transparency as students enter the classroom and find their seats.

MOTIVATE *(10 minutes)*

___ **Demonstration, The Same, But Different, TE** Students identify crystals that are the same mineral. **(GENERAL)**

TEACH *(65 minutes)*

___ **Directed Reading, CRF** These worksheets reinforce basic concepts and vocabulary presented in the section. **(BASIC)**

___ **Activity, Those Are the Breaks, TE** Students use craft sticks and plaster of paris to create a mineral model. **(GENERAL)**

KEY **SE** = Student Edition **TE** = Teacher Edition **CRF** = Chapter Resource File

Lesson Plan

__ **Reading Skill Builder, Reading Organizer, TE** Students are encouraged to take notes while reading this section. **(BASIC)**

__ **Graphic Organizer, Comparison Table, SE** Use the Graphic Organizer to help students develop their analytical skills. **(BASIC)**

__ **Teaching Transparency 23, Moh's Hardness Scale with Worksheet** Use this Teaching Transparency and the accompanying worksheet to support your classroom discussion of Moh's hardness scale.

__ **Activity, Guest Speaker, TE** Invite a rock or mineral collector to speak to the class. **(GENERAL)**

__ **Skill Builder, Writing, TE** Students write an illustrated paper describing some common mineral habits. **(ADVANCED)**

__ **Activity, Using Density to Identify Minerals, TE** Students calculate the density of a sample and then try to identify the mineral. **(GENERAL)**

__ **Teaching Transparency 24, The Six Basic Crystal Systems with Worksheet** Use this Teaching Transparency and the accompanying worksheet to support your classroom discussion of the six basic crystal systems.

__ **Discussion, Crystal Powers, TE** Students discuss crystal healing. **(GENERAL)**

__ **Quick Lab, Determining Density, SE** Students use a triple-beam balance, a graduated cylinder, and water to calculate the densities of mineral samples. **(GENERAL)**

__ **Datasheet for Quick Lab, Determining Density, CRF** Students use the datasheet to complete the Quick Lab. **(GENERAL)**

__ **Skills Practice Lab, Mineral Identification, SE** Students observe mineral samples and record their color/luster, hardness, streak, cleavage/fracture, and name. **(GENERAL)**

__ **Datasheet for Chapter Lab, Skills Practice Lab, Mineral Identification, CRF** Students use the datasheet to complete the Chapter Lab. **(GENERAL)**

__ **Activity, Finding Mineral Deposits, TE** Students research the methods used to locate and assess mineral deposits. **(GENERAL)**

__ **Mapping Expeditions, Buried Treasure, SE** Students use maps to learn about where resources are located in the U.S. and how this distribution affects population. **(GENERAL)**

__ **Inquiry Lab, Growing Crystals, CRF** Students grow crystals from salt solutions and examine the effects of temperature on crystal growth. **(GENERAL)**

__ **Interactive Tutor CD-ROM, Matter and Minerals** Students practice what they have learned using this interactive activity. **(GENERAL)**

__ **Interactive Tutor CD-ROM, Weight, Mass, Volume, and Density** Students practice what they have learned using this interactive activity. **(GENERAL)**

CLOSE *(10 minutes)*

__ **Section Review, SE** Students answer end-of-section vocabulary, key ideas, critical thinking, and concept mapping questions. **(GENERAL)**

KEY **SE** = Student Edition **TE** = Teacher Edition **CRF** = Chapter Resource File

Lesson Plan

__ **Reteaching, Peer Reviewing, TE** In pairs, students compose questions and quiz each other on the material in this section. **(BASIC)**

__ **Quiz, TE** Students answer 3 questions about this section. **(GENERAL)**

__ **Section Quiz, CRF** Use this quiz to assess students' understanding of the section. **(GENERAL) Also in Spanish**

__ **Spanish Section Quiz, Spanish Assessments** Use this quiz to assess students' understanding of the section. **(GENERAL)**

__ **Alternative Assessment, Crystal Models, TE** Students build models of the six crystal shapes to understand their symmetrical features. **(GENERAL)**

KEY **SE** = Student Edition **TE** = Teacher Edition **CRF** = Chapter Resource File

End of Chapter Review and Assessment

PACING

Regular Schedule: **with lab(s): N/A** **without lab(s): 2 days**
Block Schedule: **with lab(s): N/A** **without lab(s): 1 day**

__ **Chapter Review, SE** Students answer end-of-chapter vocabulary, key ideas, critical thinking, math, writing, and interpreting graphics questions. **(GENERAL)**

__ **Alternative Assessment, Mineral Displays, TE** Students create posters that display three of the five common minerals in their assigned class. **(ADVANCED)**

__ **Standardized Test Prep, SE** Have students complete the standardized test prep to help them prepare for standardized tests. **(GENERAL)**

__ **Holt PuzzlePro®** Use this software to create crossword puzzles and word searches that make learning vocabulary fun. **(GENERAL)**

__ **Concept Review, CRF** Use this worksheet to assess student's understanding of the main concepts presented in the chapter. **(GENERAL)**

__ **Critical Thinking, CRF** Students use this worksheet to develop and practice critical thinking skills. **(ADVANCED)**

__ **Math Skills, CRF** This worksheet helps reinforce students' understanding of geometry. **(GENERAL)**

__ **Graphing Skills, CRF** This worksheet helps reinforce students' understanding of ternary diagrams. **(GENERAL)**

__ **Concept Review, Study Guide, Minerals of Earth's Crust** Use this worksheet to assess student's understanding of the main concepts presented in the chapter. **(GENERAL)**

__ **Concept Review, Spanish Study Guide, Minerals of Earth's Crust** This worksheet helps reinforce English-language learners' understanding of the main concepts presented in the chapter. **(GENERAL)**

__ **Chapter Test A, CRF** Assign questions from the appropriate test for chapter assessment. **(GENERAL)**

__ **Chapter Test B, CRF** Assign questions from the appropriate test for chapter assessment. **(ADVANCED)**

__ **Spanish Chapter Test A, Spanish Assessments** Assign questions from the appropriate test for chapter assessment. **(GENERAL)**

__ **Test Generator, One-Stop Planner** Create a customized homework assignment, quiz, or test using the HRW Test Generator program. **(GENERAL)**

__ **Test Item Listing, CRF** Use the Test Item Listing to identify questions for a customized homework, quiz, or test.

__ **PowerPoint® Resources** Use the customizable presentation to help students practice standardized test-taking skills.

KEY **SE** = Student Edition **TE** = Teacher Edition **CRF** = Chapter Resource File

OTHER RESOURCE OPTIONS

__ **HRW Earth Science Video, Segment 8, "Minerals of the Earth's Crust"** This segment focuses on specific properties of minerals, which can often be used to identify mineral species. **(GENERAL)**

__ **Brain Food Video Quiz** Use this video resource to assess student's progress. **(GENERAL)**

KEY **SE** = Student Edition **TE** = Teacher Edition **CRF** = Chapter Resource File

Chapter 6 Rocks

Chapter Opener

__ **Pre-Reading Activity, Pyramid, SE**

__ **Using the Figure, Eagletail Peak, TE (GENERAL)**

Section 1 Rocks and the Rock Cycle

PACING

Regular Schedule:	**with lab(s): 2 days**	**without lab(s): 1 day**
Block Schedule:	**with lab(s): 1 day**	**without lab(s): 0.5 day**

OBJECTIVES

1. **Identify** the three major types of rock, and explain how each type forms.
2. **Summarize** the steps in the rock cycle.
3. **Explain** Bowen's reaction series.
4. **Summarize** the factors that affect the stability of rocks.

NATIONAL SCIENCE EDUCATION STANDARDS

ES 2b Movement of matter between reservoirs is driven by the earth's internal and external sources of energy. These movements are often accompanied by a change in the physical and chemical properties of the matter. Carbon, for example, occurs in carbonate rocks such as limestone, in the atmosphere as carbon dioxide gas, in water as dissolved carbon dioxide, and in all organisms as complex molecules that control the chemistry of life.

FOCUS *(5 minutes)*

__ **Overview, TE** Review the objectives listed in the Student Edition. **(GENERAL)**

__ **PowerPoint® Resources** Use the customizable presentation to enhance your lecture.

__ **Bellringer, TE** Have students list the names of five rocks with which they are familiar. **(GENERAL)**

__ **Bellringer Transparency** Use this transparency as students enter the classroom and find their seats.

MOTIVATE *(10 minutes)*

__ **Discussion, Types of Rocks, TE** Students discuss the features of igneous, sedimentary, and metamorphic rock samples. **(GENERAL)**

KEY **SE** = Student Edition **TE** = Teacher Edition **CRF** = Chapter Resource File

TEACH *(65 minutes)*

__ **Directed Reading, CRF** These worksheets reinforce basic concepts and vocabulary presented in the section. **(BASIC)**

__ **Using the Figure, The Rock Cycle, TE** Students identify the processes that cause a rock to change from one type to another or into a different form of its original type. **(GENERAL)**

__ **Teaching Transparency 26, The Rock Cycle with Worksheet** Use this Teaching Transparency and the accompanying worksheet to support your classroom discussion of the rock cycle.

__ **Teaching Transparency 27, Bowen's Reaction Series with Worksheet** Use this Teaching Transparency and the accompanying worksheet to support your classroom discussion of Bowen's reaction series.

__ **Graphic Organizer, Chain-of-Events Chart, SE** Use the graphic organizer to help students develop their analytical skills. **(GENERAL)**

__ **Skills Practice Lab, Classification of Rocks, SE** Students use distinguishing rock features to identify various rock samples. **(GENERAL)**

__ **Datasheet for Chapter Lab, Skills Practice Lab, Classification of Rocks, CRF** Students use the datasheet to complete the Chapter Lab. **(GENERAL)**

__ **Maps in Action, Geologic Map of Virginia, SE** Students enhance their map-reading skills by answering questions about a map showing geologic data for the state of Virginia. **(GENERAL)**

__ **Teaching Transparency 31, Geologic Map of Virginia with Worksheet** Use this Teaching Transparency and the accompanying worksheet to help students perform the Maps in Action activity.

__ **Activity, Analyzing Rocks, SE** Students research a region to learn the age of the rocks found there and present their findings. **(GENERAL)**

__ **Science and Technology, Moon Rock, SE** The different types of moon rocks and the methods of collecting and studying them are discussed.

__ **Internet Activity, Geology of Other Planets, TE** Students research geological information about other bodies in the Solar System . **(ADVANCED)**

__ **Internet Activity, Geology of Other Planets, CRF** This worksheet directs student research for the topic presented in the Teacher Edition. **(ADVANCED)**

CLOSE *(10 minutes)*

__ **Section Review, SE** Students answer end-of-section vocabulary, key ideas, critical thinking, and concept mapping questions. **(GENERAL)**

__ **Reteaching, Rock Cycle Processes, TE** Students review the formation processes of igneous, metamorphic, and sedimentary rocks. **(BASIC)**

__ **Quiz, TE** Students answer 2 questions about this section. **(GENERAL)**

__ **Section Quiz, CRF** Use this quiz to assess students' understanding of the section. **(GENERAL) Also in Spanish**

KEY **SE** = Student Edition **TE** = Teacher Edition **CRF** = Chapter Resource File

__ **Spanish Section Quiz, Spanish Assessments** Use this quiz to assess students' understanding of the section. **(GENERAL)**

__ **Alternative Assessment, Rock Crystallization, TE** Students create 3-dimensional models of the rock cycle. **(ADVANCED)**

KEY **SE** = Student Edition **TE** = Teacher Edition **CRF** = Chapter Resource File

Lesson Plan

Section 2 Igneous Rock

PACING

Regular Schedule:	**with lab(s): N/A**	**without lab(s): 2 days**
Block Schedule:	**with lab(s): N/A**	**without lab(s): 1 day**

OBJECTIVES

1. **Summarize** three factors that affect whether rock melts.
2. **Describe** how the cooling rate of magma and lava affects the texture of igneous rocks.
3. **Classify** igneous rocks according to their composition and texture.
4. **Describe** intrusive and extrusive igneous rock structures.

NATIONAL SCIENCE EDUCATION STANDARDS

ES 2b Movement of matter between reservoirs is driven by the earth's internal and external sources of energy. These movements are often accompanied by a change in the physical and chemical properties of the matter. Carbon, for example, occurs in carbonate rocks such as limestone, in the atmosphere as carbon dioxide gas, in water as dissolved carbon dioxide, and in all organisms as complex molecules that control the chemistry of life.

FOCUS *(5 minutes)*

__ **Overview, TE** Review the objectives listed in the Student Edition. **(GENERAL)**

__ **PowerPoint® Resources** Use the customizable presentation to enhance your lecture.

__ **Bellringer, TE** Have students to write a short paragraph that explains where and why rock melts. **(GENERAL)**

__ **Bellringer Transparency** Use this transparency as students enter the classroom and find their seats.

MOTIVATE *(10 minutes)*

__ **Discussion, Hand Samples, TE** Students compare igneous rock samples. **(GENERAL)**

TEACH *(20 minutes)*

__ **Directed Reading, CRF** These worksheets reinforce basic concepts and vocabulary presented in the section. **(BASIC)**

__ **Teaching Transparency 28, Partial Melting and Fractional Crystallization with Worksheet** Use this Teaching Transparency and the accompanying worksheet to support your classroom discussion of partial melting and fractional crystallization.

KEY **SE** = Student Edition **TE** = Teacher Edition **CRF** = Chapter Resource File

__ **Quick Lab, Crystal Formation, SE** Students conduct an experiment to study crystallization. **(GENERAL)**

__ **Datasheet for Quick Lab, Crystal Formation, CRF** Students use the datasheet to complete the Quick Lab. **(GENERAL)**

__ **Using the Figure, Fractional Crystallization, TE** Students study the figure showing crystallization from magma and compare it to partial melting. **(BASIC)**

__ **Demonstration, Partial Freezing, TE** Students observe a demonstration of fractional crystallization. **(GENERAL)**

__ **Reading Skill Builder, Reading Organizer, TE** Students review descriptions of igneous rocks. **(BASIC)**

__ **Geology Connection, Information from Igneous Rocks, TE** Students research where igneous rocks can be found and what geological clues they provide. **(ADVANCED)**

__ **Group Activity, Poster Project, TE** Students work in groups to create a poster that lists properties and examples of felsic, mafic, and intermediate rocks. **(ADVANCED)**

__ **Using the Figure, Intrusive Igneous Structures, TE** Students use the figure showing igneous intrusions to describe how the intrusions caused landforms. **(GENERAL)**

__ **Skill Builder, Vocabulary, TE** Students review the origins of terms for intrusive igneous structures. **(GENERAL)**

CLOSE *(10 minutes)*

__ **Section Review, SE** Students answer end-of-section vocabulary, key ideas, critical thinking, and concept mapping questions. **(GENERAL)**

__ **Reteaching, Igneous Rock Characteristics, TE** Students create a table using the names of igneous rocks. **(BASIC)**

__ **Quiz, TE** Students answer 3 questions about this section. **(GENERAL)**

__ **Section Quiz, CRF** Use this quiz to assess students' understanding of the section. **(GENERAL) Also in Spanish**

__ **Spanish Section Quiz, Spanish Assessments** Use this quiz to assess students' understanding of the section. **(GENERAL)**

__ **Alternative Assessment, Travel Brochures, TE** Students create travel brochures for 5 locations that have exposed intrusive or extrusive igneous structures. **(ADVANCED)**

KEY **SE** = Student Edition **TE** = Teacher Edition **CRF** = Chapter Resource File

Lesson Plan

Section 3 Sedimentary Rock

PACING

Regular Schedule: **with lab(s): N/A** **without lab(s): 2 days**
Block Schedule: **with lab(s): N/A** **without lab(s): 1 day**

OBJECTIVES

1. **Explain** the processes of compaction and cementation.
2. **Describe** how chemical and organic sedimentary rocks form.
3. **Describe** how clastic sedimentary rock forms.
4. **Identify** seven sedimentary rock features.

NATIONAL SCIENCE EDUCATION STANDARDS

ES 2b Movement of matter between reservoirs is driven by the earth's internal and external sources of energy. These movements are often accompanied by a change in the physical and chemical properties of the matter. Carbon, for example, occurs in carbonate rocks such as limestone, in the atmosphere as carbon dioxide gas, in water as dissolved carbon dioxide, and in all organisms as complex molecules that control the chemistry of life.

FOCUS *(5 minutes)*

__ **Overview, TE** Review the objectives listed in the Student Edition. **(GENERAL)**

__ **PowerPoint® Resources** Use the customizable presentation to enhance your lecture.

__ **Bellringer, TE** Have students to write down three features that they associate with sedimentary rock compared with other kinds of rock. **(GENERAL)**

__ **Bellringer Transparency** Use this transparency as students enter the classroom and find their seats.

MOTIVATE *(10 minutes)*

__ **Demonstration, Sedimentation, TE** Students observe a demonstration of the basic process of sedimentation. **(GENERAL)**

TEACH *(20 minutes)*

__ **Directed Reading, CRF** These worksheets reinforce basic concepts and vocabulary presented in the section. **(BASIC)**

__ **Teaching Transparency 29, Organic Limestone Formation with Worksheet** Use this Teaching Transparency and the accompanying worksheet to support your classroom discussion of organic limestone formation.

KEY **SE** = Student Edition **TE** = Teacher Edition **CRF** = Chapter Resource File

__ **Using the Figure, Organic Limestone Formation, TE** Students use the figures showing organic limestone formation to answer questions about limestone. **(GENERAL)**

__ **Chemistry Connection, The Carbon Cycle, TE** Students research various steps of the carbon cycle and present their findings. **(ADVANCED)**

__ **Discussion, Comprehension Check, TE** Students discuss clastic sedimentary rock. **(GENERAL)**

__ **Reading Skill Builder, Paired Summarizing, TE** Students read the text in pairs and take turns summarizing what they have read. **(BASIC)**

__ **Internet Activity, Identifying Sedimentary Rock Features, TE** Students research examples of sedimentary rock features and present their findings. **(ADVANCED)**

__ **Internet Activity, Identifying Sedimentary Rock Features, CRF** This worksheet directs student research for the topic presented in the Teacher Edition. **(ADVANCED)**

__ **Reading Skill Builder, Reading Organizer, TE** Students are encouraged to take notes while reading this section. **(BASIC)**

__ **Quick Lab, Graded Bedding, SE** Students use water, a glass jar with a lid, and sediment to learn about graded bedding. **(GENERAL)**

__ **Datasheet for Quick Lab, Graded Bedding, CRF** Students use the datasheet to complete the Quick Lab. **(GENERAL)**

__ **Inquiry Lab, Sorting Sediments, CRF** Students examine how streams of water with varying amounts of force affect the sorting of sediments. **(GENERAL)**

CLOSE *(10 minutes)*

__ **Section Review, SE** Students answer end-of-section vocabulary, key ideas, critical thinking, and concept mapping questions. **(GENERAL)**

__ **Homework, Chemical Sedimentary Rock, TE** Students research different types of chemical sedimentary rock. **(GENERAL)**

__ **Reteaching, Sedimentary Rocks, TE** Students match terms relating to sedimentary rocks. **(BASIC)**

__ **Quiz, TE** Students answer 2 questions about this section. **(GENERAL)**

__ **Section Quiz, CRF** Use this quiz to assess students' understanding of the section. **(GENERAL) Also in Spanish**

__ **Spanish Section Quiz, Spanish Assessments** Use this quiz to assess students' understanding of the section. **(GENERAL)**

__ **Alternative Assessment, Modeling Sedimentation, TE** Students create models showing the various stages in the formation of sedimentary rock. **(ADVANCED)**

KEY **SE** = Student Edition **TE** = Teacher Edition **CRF** = Chapter Resource File

Lesson Plan

Section 4 Metamorphic Rock

PACING

Regular Schedule:	**with lab(s): N/A**	**without lab(s): 2 days**
Block Schedule:	**with lab(s): N/A**	**without lab(s): 1 day**

OBJECTIVES

1. **Describe** the process of metamorphism.
2. **Explain** the difference between regional and contact metamorphism.
3. **Distinguish** between foliated and nonfoliated metamorphic rocks, and give an example of each.

NATIONAL SCIENCE EDUCATION STANDARDS

ES 2b Movement of matter between reservoirs is driven by the earth's internal and external sources of energy. These movements are often accompanied by a change in the physical and chemical properties of the matter. Carbon, for example, occurs in carbonate rocks such as limestone, in the atmosphere as carbon dioxide gas, in water as dissolved carbon dioxide, and in all organisms as complex molecules that control the chemistry of life.

FOCUS *(5 minutes)*

__ **Overview, TE** Review the objectives listed in the Student Edition. **(GENERAL)**

__ **PowerPoint® Resources** Use the customizable presentation to enhance your lecture.

__ **Bellringer, TE** Ask students to list items that change from one form to another. **(GENERAL)**

__ **Bellringer Transparency** Use this transparency as students enter the classroom and find their seats.

MOTIVATE *(10 minutes)*

__ **Demonstration, Metamorphism, TE** Students observe similarities and differences in rock samples. **(GENERAL)**

TEACH *(20 minutes)*

__ **Directed Reading, CRF** These worksheets reinforce basic concepts and vocabulary presented in the section. **(BASIC)**

__ **Teaching Transparency 30, Indicators of Metamorphic Conditions with Worksheet** Use this Teaching Transparency and the accompanying worksheet to support your classroom discussion of indicators of metamorphic conditions.

KEY **SE** = Student Edition **TE** = Teacher Edition **CRF** = Chapter Resource File

__ **Reading Skill Builder, Anticipation Guide, TE** Students predict the difference between contact metamorphism and regional metamorphism before reading the page. **(BASIC)**

__ **Activity, Poster Project, TE** Students research an area where contact or regional metamorphism is known to occur and present their findings. **(ADVANCED)**

__ **Interactive Tutor CD-ROM, Igneous, Sedimentary, and Metamorphic Rocks** Students practice what they have learned using this interactive activity. **(GENERAL)**

__ **Making Models Lab, Metamorphic Rocks, CRF** Students model metamorphism to see how pressure reorients mineral grains in metamorphic rock. **(GENERAL)**

CLOSE *(10 minutes)*

__ **Section Review, SE** Students answer end-of-section vocabulary, key ideas, critical thinking, and concept mapping questions. **(GENERAL)**

__ **Homework, Degrees of Metamorphism, TE** Students research metamorphic grades of rock. **(ADVANCED)**

__ **Reteaching, Metamorphic Rocks, TE** Students review examples of foliated and nonfoliated rocks. **(BASIC)**

__ **Quiz, TE** Students answer 3 questions about this section. **(GENERAL)**

__ **Section Quiz, CRF** Use this quiz to assess students' understanding of the section. **(GENERAL) Also in Spanish**

__ **Spanish Section Quiz, Spanish Assessments** Use this quiz to assess students' understanding of the section. **(GENERAL)**

__ **Alternative Assessment, Metamorphic Chemistry, TE** Students research the geological and chemical processes that form one type of gemstone. **(ADVANCED)**

KEY **SE** = Student Edition **TE** = Teacher Edition **CRF** = Chapter Resource File

End of Chapter Review and Assessment

PACING

Regular Schedule: **with lab(s): N/A** **without lab(s): 2 days**
Block Schedule: **with lab(s): N/A** **without lab(s): 1 day**

__ **Chapter Review, SE** Students answer end-of-chapter vocabulary, key ideas, critical thinking, math, writing, and interpreting graphics questions. **(GENERAL)**

__ **Alternative Assessment, Bulletin Board Display, TE** Students create a display showing the processes by which rocks form and transform into other rocks. **(GENERAL)**

__ **Standardized Test Prep, SE** Have students complete the standardized test prep to help them prepare for standardized tests. **(GENERAL)**

__ **Holt PuzzlePro®** Use this software to create crossword puzzles and word searches that make learning vocabulary fun. **(GENERAL)**

__ **Concept Review, CRF** Use this worksheet to assess student's understanding of the main concepts presented in the chapter. **(GENERAL)**

__ **Critical Thinking, CRF** Students use this worksheet to develop and practice critical thinking skills. **(ADVANCED)**

__ **Math Skills, CRF** This worksheet helps reinforce students' understanding of significant digits. **(GENERAL)**

__ **Graphing Skills, CRF** This worksheet helps reinforce students' understanding of ternary diagrams. **(GENERAL)**

__ **Concept Review, Study Guide, Rocks** Use this worksheet to assess student's understanding of the main concepts presented in the chapter. **(GENERAL)**

__ **Concept Review, Spanish Study Guide, Rocks** This worksheet helps reinforce English-language learners' understanding of the main concepts presented in the chapter. **(GENERAL)**

__ **Chapter Test A, CRF** Assign questions from the appropriate test for chapter assessment. **(GENERAL)**

__ **Chapter Test B, CRF** Assign questions from the appropriate test for chapter assessment. **(ADVANCED)**

__ **Spanish Chapter Test A, Spanish Assessments** Assign questions from the appropriate test for chapter assessment. **(GENERAL)**

__ **Test Generator, One-Stop Planner** Create a customized homework assignment, quiz, or test using the HRW Test Generator program. **(GENERAL)**

__ **Test Item Listing, CRF** Use the Test Item Listing to identify questions for a customized homework, quiz, or test. **(GENERAL)**

__ **PowerPoint® Resources** Use the customizable presentation to help students practice standardized test-taking skills. **(GENERAL)**

KEY **SE** = Student Edition **TE** = Teacher Edition **CRF** = Chapter Resource File

OTHER RESOURCE OPTIONS

__ **Brain Food Video Quiz** Use this video resource to assess student's progress. **(GENERAL)**

KEY **SE** = Student Edition **TE** = Teacher Edition **CRF** = Chapter Resource File

Chapter 7 Resources and Energy

Chapter Opener

__ **Pre-Reading Activity, Four-Corner Fold, SE**

__ **Using the Figure, Hoover Dam, TE (GENERAL)**

__ **Long-Term Project, Air Pollution Watch, SE** Students collect particulates for two weeks and examine how wind direction affects their numbers. **(ADVANCED)**

__ **Long-Term Project, Air Pollution Watch, Long-Term Projects** Students use this datasheet to complete the Long-Term Project. **(ADVANCED)**

Section 1 Mineral Resources

PACING

Regular Schedule:	**with lab(s): N/A**	**without lab(s): 2 days**
Block Schedule:	**with lab(s): N/A**	**without lab(s): 1 day**

OBJECTIVES

1. **Explain** what ores are and how they form.
2. **Identify** four uses for mineral resources.
3. **Summarize** two ways humans obtain mineral resources.

NATIONAL SCIENCE EDUCATION STANDARDS

PS 3a Chemical reactions occur all around us, for example in health care, cooking, cosmetics, and automobiles. Complex chemical reactions involving carbon-based molecules take place constantly in every cell in our bodies.

PS 3b Chemical reactions may release or consume energy. Some reactions such as the burning of fossil fuels release large amounts of energy by losing heat and by emitting light. Light can initiate many chemical reactions such as photosynthesis and the evolution of urban smog.

SPSP 3a Human populations use resources in the environment in order to maintain and improve their existence. Natural resources have been and will continue to be used to maintain human populations.

FOCUS *(5 minutes)*

__ **Overview, TE** Review the objectives listed in the Student Edition. **(GENERAL)**

__ **PowerPoint® Resources** Use the customizable presentation to enhance your lecture.

__ **Bellringer, TE** Ask students to define the word *mineral,* and then name 5 common minerals. **(GENERAL)**

KEY **SE** = Student Edition **TE** = Teacher Edition **CRF** = Chapter Resource File

Lesson Plan

__ **Bellringer Transparency** Use this transparency as students enter the classroom and find their seats.

MOTIVATE *(10 minutes)*

__ **Group Activity, Minerals at Work, TE** Students list common objects that are made from minerals, and then identify their characteristics. **(GENERAL)**

TEACH *(20 minutes)*

__ **Directed Reading, CRF** These worksheets reinforce basic concepts and vocabulary presented in the section. **(BASIC)**

__ **Teaching Transparency 32, The Formation of Ores and Placer Deposits with Worksheet** Use this Teaching Transparency and the accompanying worksheet to support your classroom discussion of the formation of ores and placer deposits.

__ **Using the Figure, Slowing Water Flow, TE** Students identify locations of ore deposits and relate those locations to the stream features shown in the figure. **(GENERAL)**

__ **Reading Skill Builder, Reading Organizer, TE** Students create charts showing the chemical abbreviations of elements discussed in this section. **(BASIC)**

__ **Discussion, Properties of Minerals, TE** Students review the different methods used in mineral exploration. **(ADVANCED)**

__ **Group Activity, Mineral Use, TE** In groups, students research the specific minerals and gemstones that are used in everyday objects. **(GENERAL)**

__ **Eye on the Environment, What's Mined is Yours, SE** The effects of mining and the costs of reclamation are discussed.

__ **Group Activity, Modern Mining, TE** In groups, students research how a particular resource is mined and create presentations to share their findings. **(BASIC)**

__ **Mapping Expeditions, Buried Treasure, SE** Students study maps showing the distribution of resources and population in the United States. **(GENERAL)**

__ **Skills Practice Lab, Determination of Carbonate in Ore, CRF** Students use chemical reactions to calculate the mass of carbonate in soil samples. **(GENERAL)**

CLOSE *(10 minutes)*

__ **Section Review, SE** Students answer end-of-section vocabulary, key ideas, critical thinking, and concept mapping questions. **(GENERAL)**

__ **Reteaching, Ore Minerals, TE** Students make tables identifying common metals and minerals, how they are used, and how they are mined. **(BASIC)**

__ **Quiz, TE** Students answer 2 questions about this section. **(GENERAL)**

__ **Section Quiz, CRF** Use this quiz to assess students' understanding of the section. **(GENERAL) Also in Spanish**

__ **Spanish Section Quiz, Spanish Assessments** Use this quiz to assess students' understanding of the section. **(GENERAL)**

KEY **SE** = Student Edition **TE** = Teacher Edition **CRF** = Chapter Resource File

__ **Alternative Assessment, Chart, TE** Students create flowcharts showing how a mineral becomes part of a deposit, is mined, and is used to make an object. **(GENERAL)**

KEY **SE** = Student Edition **TE** = Teacher Edition **CRF** = Chapter Resource File

Section 2 Nonrenewable Energy

PACING

Regular Schedule:	**with lab(s): N/A**	**without lab(s): 2 days**
Block Schedule:	**with lab(s): N/A**	**without lab(s): 1 day**

OBJECTIVES

1. **Explain** why coal is a fossil fuel.
2. **Describe** how petroleum and natural gas form and how they are removed from Earth.
3. **Summarize** the processes of nuclear fission and nuclear fusion.
4. **Explain** how nuclear fission generates electricity.

NATIONAL SCIENCE EDUCATION STANDARDS

LS 4a The atoms and molecules on the earth cycle among the living and nonliving components of the biosphere.

PS 1c The nuclear forces that hold the nucleus of an atom together, at nuclear distances, are usually stronger than the electric forces that would make it fly apart. Nuclear reactions convert a fraction of the mass of interacting particles into energy, and they can release much greater amounts of energy than atomic interactions. Fission is the splitting of a large nucleus into smaller pieces. Fusion is the joining of two nuclei at extremely high temperature and pressure, and is the process responsible for the energy of the sun and other stars.

PS 2f Carbon atoms can bond to one another in chains, rings, and branching networks to form a variety of structures, including synthetic polymers, oils, and the large molecules essential to life.

PS 3a Chemical reactions occur all around us, for example in health care, cooking, cosmetics, and automobiles. Complex chemical reactions involving carbon-based molecules take place constantly in every cell in our bodies.

PS 3c A large number of important reactions involve the transfer of either electrons (oxidation/reduction reactions) or hydrogen ions (acid/base reactions) between reacting ions, molecules, or atoms. In other reactions, chemical bonds are broken by heat or light to form very reactive radicals with electrons ready to form new bonds. Radical reactions control many processes such as the presence of ozone and greenhouse gases in the atmosphere, burning and processing of fossil fuels, the formation of polymers, and explosions.

SPSP 3a Human populations use resources in the environment in order to maintain and improve their existence. Natural resources have been and will continue to be used to maintain human populations.

KEY **SE** = Student Edition **TE** = Teacher Edition **CRF** = Chapter Resource File

Lesson Plan

FOCUS *(5 minutes)*

__ **Overview, TE** Review the objectives listed in the Student Edition. **(GENERAL)**

__ **PowerPoint® Resources** Use the customizable presentation to enhance your lecture.

__ **Bellringer, TE** Ask students: "What would your life be like without electricity or gasoline?" **(GENERAL)**

__ **Bellringer Transparency** Use this transparency as students enter the classroom and find their seats.

MOTIVATE *(10 minutes)*

__ **Identifying Preconceptions, Is It Renewable? TE** Students review a list of resources and identify which ones are renewable. **(GENERAL)**

TEACH *(20 minutes)*

__ **Directed Reading, CRF** These worksheets reinforce basic concepts and vocabulary presented in the section. **(BASIC)**

__ **Group Activity, Purifying with Pressure, TE** In groups, students use weights and wet towels to see how applied pressure removes impurities from compressed substances. **(GENERAL)**

__ **Internet Activity, Resource Locations, TE** Students use the internet to search for maps showing the locations of coal deposits on Earth. **(GENERAL)**

__ **Internet Activity, Resource Locations, CRF** This worksheet directs student research for the topic presented in the Teacher Edition. **(GENERAL)**

__ **Teaching Transparency 33, Types of Coal with Worksheet** Use this Teaching Transparency and the accompanying worksheet to support your classroom discussion of types of coal.

__ **Using the Figure, Rock Layers and Density, TE** Students study the figures of oil traps and describe characteristics of the rock layers. **(GENERAL)**

__ **Teaching Transparency 34, Oil Traps with Worksheet** Use this Teaching Transparency and the accompanying worksheet to support your classroom discussion of oil traps.

__ **Demonstration, Chain Reaction, TE** Students stand up and shake one person's hand to see how chain reactions work. **(BASIC)**

__ **Skill Builder, Vocabulary, TE** Students discuss the origins of the words *nucleus* and *nuclear*. **(GENERAL)**

__ **Teaching Transparency 35, A Nuclear Fission Reaction with Worksheet** Use this Teaching Transparency and the accompanying worksheet to support your classroom discussion of nuclear fission reactions.

__ **Teaching Transparency 36, How a Nuclear Power Plant Generates Electricity with Worksheet** Use this Teaching Transparency and the accompanying worksheet to support your classroom discussion of how nuclear power plants generate electricity.

KEY SE = Student Edition TE = Teacher Edition CRF = Chapter Resource File

__ **History Connection, Nuclear Theory, TE** Students research scientists whose work has led to the development of nuclear theory. **(ADVANCED)**

__ **Graphic Organizer, Chain-of-Events Chart, SE** Use the Graphic Organizer to help students develop their analytical skills. **(GENERAL)**

__ **Using the Figure, Electricity from Nuclear Fission, TE** Students discuss why fuel rods are housed in such large, reinforced structures. **(GENERAL)**

__ **Debate, Pros and Cons, TE** Students debate whether or not the risks of nuclear power are acceptable for creating electricity. **(GENERAL)**

__ **Interactive Tutor CD-ROM, What's the Matter? (Atomic Theory and Nuclear Energy)** Students practice what they have learned using this interactive activity. **(GENERAL)**

CLOSE *(10 minutes)*

__ **Section Review, SE** Students answer end-of-section vocabulary, key ideas, critical thinking, and concept mapping questions. **(GENERAL)**

__ **Homework, Fudging Fusion? TE** Students research cold fusion. **(ADVANCED)**

__ **Reteaching, Flashcards, TE** Students make flashcards of energy sources, then work in pairs to reinforce the content of this section. **(BASIC)**

__ **Quiz, TE** Students answer 2 questions about this section. **(GENERAL)**

__ **Section Quiz, CRF** Use this quiz to assess students' understanding of the section. **(GENERAL) Also in Spanish**

__ **Spanish Section Quiz, Spanish Assessments** Use this quiz to assess students' understanding of the section. **(GENERAL)**

__ **Alternative Assessment, Letter to the Editor, TE** Students write letters to the editor of a local paper in which they advocate the use of either fossil fuels or nuclear energy. **(GENERAL)**

KEY **SE** = Student Edition **TE** = Teacher Edition **CRF** = Chapter Resource File

Section 3 Renewable Energy

PACING

Regular Schedule: **with lab(s): 2 days** **without lab(s): 1 day**
Block Schedule: **with lab(s): 1 day** **without lab(s): 0.5 day**

OBJECTIVES

1. **Explain** how geothermal energy may be used as a substitute for fossil fuels.
2. **Compare** passive and active methods of harnessing energy from the sun.
3. **Explain** how water and wind can be harnessed to generate electricity.

NATIONAL SCIENCE EDUCATION STANDARDS

PS 3a Chemical reactions occur all around us, for example in health care, cooking, cosmetics, and automobiles. Complex chemical reactions involving carbon-based molecules take place constantly in every cell in our bodies.

PS 3b Chemical reactions may release or consume energy. Some reactions such as the burning of fossil fuels release large amounts of energy by losing heat and by emitting light. Light can initiate many chemical reactions such as photosynthesis and the evolution of urban smog.

SPSP 3a Human populations use resources in the environment in order to maintain and improve their existence. Natural resources have been and will continue to be used to maintain human populations.

FOCUS *(5 minutes)*

__ **Overview, TE** Review the objectives listed in the Student Edition. **(GENERAL)**

__ **PowerPoint® Resources** Use the customizable presentation to enhance your lecture.

__ **Bellringer, TE** Ask students to list things that they can renew, and then have them come up with a definition of *renewable*. **(GENERAL)**

__ **Bellringer Transparency** Use this transparency as students enter the classroom and find their seats.

MOTIVATE *(10 minutes)*

__ **Using the Figure, Natural Hot Tub, TE** Students study the figure and explain how they can tell that the water in the photo is hot. **(GENERAL)**

TEACH *(65 minutes)*

__ **Directed Reading, CRF** These worksheets reinforce basic concepts and vocabulary presented in the section. **(BASIC)**

__ **Skill Builder, Vocabulary, TE** Students learn the meaning of the prefix *photo*. **(BASIC)**

KEY **SE** = Student Edition **TE** = Teacher Edition **CRF** = Chapter Resource File

__ **Quick Lab, Solar Collector, SE** Students use pans and thermometers to study and record the rate of temperature change in their solar collectors. **(GENERAL)**

__ **Datasheet for Quick Lab, Solar Collector, CRF** Students use the datasheet to complete the Quick Lab. **(GENERAL)**

__ **Internet Activity, Biomass Potential, TE** Students use the internet to discover how much organic waste Americans generate and propose ways to use it to generate power instead. **(ADVANCED)**

__ **Internet Activity, Biomass Potential, CRF** This worksheet directs student research for the topic presented in the Teacher Edition. **(ADVANCED)**

__ **Teaching Transparency 37, How a Hydroelectric Dam Generates Electricity with Worksheet** Use this Teaching Transparency and the accompanying worksheet to support your classroom discussion of how a hydroelectric dam generates electricity.

__ **Inquiry Lab, Blowing in the Wind, SE** Students design and build a windmill. **(GENERAL)**

__ **Datasheet for Chapter Lab, Inquiry Lab, Blowing in the Wind, CRF** Students use the datasheet to complete the Chapter Lab. **(GENERAL)**

__ **Maps in Action, Wind Power in the United States, SE** Students enhance their map-reading skills by answering questions about a map showing the locations and speeds of wind power projects throughout the U.S. **(GENERAL)**

__ **Teaching Transparency 38, Wind Power in the United States with Worksheet** Use this Teaching Transparency and the accompanying worksheet to help students perform the Maps in Action activity.

__ **Internet Activity, Alternative Energy in Your State, TE** Students research alternative energy used in your state and write letters to the organization requesting more information. **(GENERAL)**

__ **Internet Activity, Alternative Energy in Your State, CRF** This worksheet directs student research for the topic presented in the Teacher Edition. **(GENERAL)**

__ **Inquiry Lab, The Generation of Natural Gas from Biomass, CRF** Students build a waste material digester to produce and gather methane gas. **(GENERAL)**

__ **Interactive Tutor CD-ROM, Renewable and Nonrenewable Resources** Students practice what they have learned using this interactive activity. **(GENERAL)**

__ **Interactive Tutor CD-ROM, Energy** Students practice what they have learned using this interactive activity. **(GENERAL)**

CLOSE *(10 minutes)*

__ **Section Review, SE** Students answer end-of-section vocabulary, key ideas, critical thinking, and concept mapping questions. **(GENERAL)**

__ **Reteaching, Energy in Action, TE** Students perform various activities demonstrating types of renewable energy. **(BASIC)**

__ **Quiz, TE** Students answer 2 questions about this section. **(GENERAL)**

KEY **SE** = Student Edition **TE** = Teacher Edition **CRF** = Chapter Resource File

__ **Section Quiz, CRF** Use this quiz to assess students' understanding of the section. **(GENERAL) Also in Spanish**

__ **Spanish Section Quiz, Spanish Assessments** Use this quiz to assess students' understanding of the section. **(GENERAL)**

__ **Alternative Assessment, Tour Brochure, TE** Students create brochures for alternative energy use in different parts of the U.S. **(GENERAL)**

KEY **SE** = Student Edition **TE** = Teacher Edition **CRF** = Chapter Resource File

Lesson Plan

Section 4 Resources and Conservation

PACING

Regular Schedule:	**with lab(s): N/A**	**without lab(s): 2 days**
Block Schedule:	**with lab(s): N/A**	**without lab(s): 1 day**

OBJECTIVES

1. **Describe** two environmental impacts of mining and the use of fossil fuels.
2. **Explain** two ways the environmental impacts of mining can be reduced.
3. **Identify** three ways that you can conserve natural resources.

NATIONAL SCIENCE EDUCATION STANDARDS

PS 3c A large number of important reactions involve the transfer of either electrons (oxidation/reduction reactions) or hydrogen ions (acid/base reactions) between reacting ions, molecules, or atoms. In other reactions, chemical bonds are broken by heat or light to form very reactive radicals with electrons ready to form new bonds. Radical reactions control many processes such as the presence of ozone and greenhouse gases in the atmosphere, burning and processing of fossil fuels, the formation of polymers, and explosions.

LS 4e Human beings live within the world's ecosystems. Increasingly, humans modify ecosystems as a result of population growth, technology, and consumption. Human destruction of habitats through direct harvesting, pollution, atmospheric changes, and other factors is threatening current global stability, and if not addressed, ecosystems will be irreversibly affected.

SPSP 3a Human populations use resources in the environment in order to maintain and improve their existence. Natural resources have been and will continue to be used to maintain human populations.

FOCUS *(5 minutes)*

__ **Overview, TE** Review the objectives listed in the Student Edition. **(GENERAL)**

__ **PowerPoint® Resources** Use the customizable presentation to enhance your lecture.

__ **Bellringer, TE** Ask students to write a paragraph that explains which renewable resource they think should be developed in case a growing population makes it necessary. **(GENERAL)**

__ **Bellringer Transparency** Use this transparency as students enter the classroom and find their seats.

KEY **SE** = Student Edition **TE** = Teacher Edition **CRF** = Chapter Resource File

MOTIVATE *(10 minutes)*

__ **Discussion, Mining, TE** Students discuss the effects of surface mining. **(GENERAL)**

TEACH *(20 minutes)*

__ **Directed Reading, CRF** These worksheets reinforce basic concepts and vocabulary presented in the section. **(BASIC)**

__ **Quick Lab, Reclamation, SE** Students use multi-layered gelatin dessert cups to demonstrate the effects of mining and reclamation. **(GENERAL)**

__ **Datasheet for Quick Lab, Reclamation, CRF** Students use the datasheet to complete the Quick Lab. **(GENERAL)**

__ **Reading Skill Builder, Reading Hint, TE** Students study the prefixes and roots of the words *conserve* and *recycle.* **(BASIC)**

__ **Group Activity, Savings and Recycling, TE** In groups, students research the energy and monetary savings that accrue due to the recycling of one product. **(GENERAL)**

__ **Using the Figure, Raw Recyclables, TE** Students discuss what processes the metal cubes in the picture must go through to make them into useable products. **(GENERAL)**

CLOSE *(10 minutes)*

__ **Section Review, SE** Students answer end-of-section vocabulary, key ideas, critical thinking, and concept mapping questions. **(GENERAL)**

__ **Reteaching, Materials We Use, TE** Students collect pictures of materials discussed in this section and create posters categorizing those pictures. **(BASIC)**

__ **Quiz, TE** Students answer 2 questions about this section. **(GENERAL)**

__ **Section Quiz, CRF** Use this quiz to assess students' understanding of the section. **(GENERAL) Also in Spanish**

__ **Spanish Section Quiz, Spanish Assessments** Use this quiz to assess students' understanding of the section. **(GENERAL)**

__ **Alternative Assessment, Survey, TE** Students visit local stores and list products containing recycled materials. **(GENERAL)**

KEY **SE** = Student Edition **TE** = Teacher Edition **CRF** = Chapter Resource File

End of Chapter Review and Assessment

PACING

Regular Schedule:	**with lab(s): N/A**	**without lab(s): 2 days**
Block Schedule:	**with lab(s): N/A**	**without lab(s): 1 day**

__ **Chapter Review, SE** Students answer end-of-chapter vocabulary, key ideas, critical thinking, math, writing, and interpreting graphics questions. **(GENERAL)**

__ **Alternative Assessment, Keeping Track, TE** Students create charts listing their energy and materials use throughout the day. **(GENERAL)**

__ **Standardized Test Prep, SE** Have students complete the standardized test prep to help them prepare for standardized tests. **(GENERAL)**

__ **Holt PuzzlePro®** Use this software to create crossword puzzles and word searches that make learning vocabulary fun. **(GENERAL)**

__ **Concept Review, CRF** Use this worksheet to assess student's understanding of the main concepts presented in the chapter. **(GENERAL)**

__ **Critical Thinking, CRF** Students use this worksheet to develop and practice critical thinking skills. **(ADVANCED)**

__ **Math Skills, CRF** This worksheet helps reinforce students' understanding of geometry. **(GENERAL)**

__ **Graphing Skills, CRF** This worksheet helps reinforce students' understanding of pie graphs. **(GENERAL)**

__ **Concept Review, Study Guide, Resources and Energy** Use this worksheet to assess student's understanding of the main concepts presented in the chapter. **(GENERAL)**

__ **Concept Review, Spanish Study Guide, Resources and Energy** This worksheet helps reinforce English-language learners' understanding of the main concepts presented in the chapter. **(GENERAL)**

__ **Chapter Test A, CRF** Assign questions from the appropriate test for chapter assessment. **(GENERAL)**

__ **Chapter Test B, CRF** Assign questions from the appropriate test for chapter assessment. **(ADVANCED)**

__ **Spanish Chapter Test A, Spanish Assessments** Assign questions from the appropriate test for chapter assessment. **(GENERAL)**

__ **Test Generator, One-Stop Planner** Create a customized homework assignment, quiz, or test using the HRW Test Generator program. **(GENERAL)**

__ **Test Item Listing, CRF** Use the Test Item Listing to identify questions for a customized homework, quiz, or test.

__ **PowerPoint® Resources** Use the customizable presentation to help students practice standardized test-taking skills.

KEY **SE** = Student Edition **TE** = Teacher Edition **CRF** = Chapter Resource File

Chapter 8 The Rock Record

Chapter Opener

__ **Pre-Reading Activity, Table Fold, SE**

__ **Using the Figure, Fossil Forms, TE (GENERAL)**

Section 1 Determining Relative Age

PACING

Regular Schedule:	**with lab(s): N/A**	**without lab(s): 2 days**
Block Schedule:	**with lab(s): N/A**	**without lab(s): 1 day**

OBJECTIVES

1. **State** the principle of uniformitarianism.
2. **Explain** how the law of superposition can be used to determine the relative age of rocks.
3. **Compare** three types of unconformities.
4. **Apply** the law of crosscutting relationships to determine the relative age of rocks.

NATIONAL SCIENCE EDUCATION STANDARDS

ES 3b Geologic time can be estimated by observing rock sequences and using fossils to correlate the sequences at various locations. Current methods include using the known decay rates of radioactive isotopes present in rocks to measure the time since the rock was formed.

HNS 3c Occasionally, there are advances in science and technology that have important and long-lasting effects on science and society. Examples of such advances include the following: Copernican revolution; Newtonian mechanics; Relativity; Geologic time scale; Plate tectonics; Atomic theory; Nuclear physics; Biological evolution; Germ theory; Industrial revolution; Molecular biology; Information and communication ; Quantum theory; Galactic universe; Medical and health technology.

UCP 4 Evolution is a series of changes, some gradual and some sporadic, that accounts for the present form and function of objects, organisms, and natural and designed systems. The general idea of evolution is that the present arises from materials and forms of the past. Although evolution is most commonly associated with the biological theory explaining the process of descent with modification of organisms from common ancestors, evolution also describes changes in the universe.

KEY **SE** = Student Edition **TE** = Teacher Edition **CRF** = Chapter Resource File

Lesson Plan

FOCUS *(5 minutes)*

__ **Overview, TE** Review the objectives listed in the Student Edition. **(GENERAL)**

__ **PowerPoint® Resources** Use the customizable presentation to enhance your lecture.

__ **Bellringer, TE** Students are asked, "What are five visual clues that help you determine if someone is older or younger than you are?" **(GENERAL)**

__ **Bellringer Transparency** Use this transparency as students enter the classroom and find their seats.

MOTIVATE *(10 minutes)*

__ **Discussion, Making Predictions, TE** Students study sedimentary rock formations and discuss which layers are the youngest and which are the oldest. **(GENERAL)**

TEACH *(20 minutes)*

__ **Directed Reading, CRF** These worksheets reinforce basic concepts and vocabulary presented in the section. **(BASIC)**

__ **Quick Lab, What's Your Relative Age? SE** In groups, students work together to arrange the members of their group from oldest to youngest. **(GENERAL)**

__ **Datasheet for Quick Lab, What's Your Relative Age? CRF** Students use the datasheet to complete the Quick Lab. **(GENERAL)**

__ **Using the Figure, The Law of Superposition, TE** Students use the figures showing rock layers to determine which layers are older and which are younger. **(GENERAL)**

__ **Discussion, Making Predictions, TE** Students predict what could disturb rock layers before reading the section on original horizontality. **(GENERAL)**

__ **Skill Builder, Vocabulary, TE** Students review the word *superposition*. **(BASIC)**

__ **Reading Skill Builder, Reading Hint, TE** Students learn to look for smaller words and word roots to help them understand the content. **(BASIC)**

__ **Teaching Transparency 39, Law of Superposition with Worksheet** Use this Teaching Transparency and the accompanying worksheet to support your classroom discussion of the law of superposition.

__ **Graphic Organizer, Spider Map, SE** Use the graphic organizer to help students develop their analytical skills. **(GENERAL)**

__ **Using the Figure, Comparing the Sedimentary Rock, TE** Students answer questions about the photos showing sedimentary rock structures. **(GENERAL)**

__ **Group Activity, Creating Layers, TE** Students use small pebbles, sand, a jar, and water to study about graded bedding in sedimentary rock. **(GENERAL)**

__ **Using the Table, Analyzing Unconformities, TE** Students use the table of unconformities to determine which rocks are older. **(ADVANCED)**

__ **Teaching Transparency 40, Types of Unconformities with Worksheet** Use this Teaching Transparency and the accompanying worksheet to support your classroom discussion of types of unconformities.

KEY **SE** = Student Edition **TE** = Teacher Edition **CRF** = Chapter Resource File

__ **Teaching Transparency 41, Crosscutting Relationships with Worksheet** Use this Teaching Transparency and the accompanying worksheet to support your classroom discussion of crosscutting relationships.

__ **Maps in Action, Geologic Map of Bedrock in Ohio, SE** Students answer questions about a map showing the ages of bedrock in Ohio to enhance their map-reading skills. **(GENERAL)**

__ **Teaching Transparency 43, Geologic Map of Bedrock in Ohio with Worksheet** Use this Teaching Transparency and the accompanying worksheet to help students perform the Maps in Action activity.

__ **Group Activity, Map Making, TE** Students work in groups to make maps of your state's bedrock geology. **(ADVANCED)**

__ **Eye on the Environment, Clues to Climate Change, SE** Evidence supporting Antarctica's climate change 60 million years ago is discussed.

__ **Internet Activity, Mapping Antarctica's Movement, TE** Students research Antarctica's movement across the planet. **(ADVANCED)**

__ **Internet Activity, Mapping Antarctica's Movement, CRF** This worksheet directs student research for the topic presented in the Teacher Edition. **(ADVANCED)**

__ **Skills Practice Lab, Determining the Relative Age of Rock Strata, CRF** Students model how forces change rock strata and practice determining relative age. **(GENERAL)**

CLOSE *(10 minutes)*

__ **Section Review, SE** Students answer end-of-section vocabulary, key ideas, critical thinking, and concept mapping questions. **(GENERAL)**

__ **Reteaching, Changing Rocks, TE** Students create models of the rock structures they learned about in this section. **(BASIC)**

__ **Quiz, TE** Students answer 2 questions about this section. **(GENERAL)**

__ **Section Quiz, CRF** Use this quiz to assess students' understanding of the section. **(GENERAL) Also in Spanish**

__ **Spanish Section Quiz, Spanish Assessments** Use this quiz to assess students' understanding of the section. **(GENERAL)**

__ **Alternative Assessment, Illustrations, TE** Students illustrate the rock structures they learned about in this section. **(GENERAL)**

KEY **SE** = Student Edition **TE** = Teacher Edition **CRF** = Chapter Resource File

Section 2 Determining Absolute Age

PACING

Regular Schedule:	**with lab(s): N/A**	**without lab(s): 2 days**
Block Schedule:	**with lab(s): N/A**	**without lab(s): 1 day**

OBJECTIVES

1. **Summarize** the limitations of using the rates of erosion and deposition to determine the absolute age of rock formations.
2. **Describe** the formation of varves.
3. **Explain** how the process of radioactive decay can be used to determine the absolute age of rocks.

NATIONAL SCIENCE EDUCATION STANDARDS

ES 3b Geologic time can be estimated by observing rock sequences and using fossils to correlate the sequences at various locations. Current methods include using the known decay rates of radioactive isotopes present in rocks to measure the time since the rock was formed.

HNS 3c Occasionally, there are advances in science and technology that have important and long-lasting effects on science and society. Examples of such advances include the following: Copernican revolution; Newtonian mechanics; Relativity; Geologic time scale; Plate tectonics; Atomic theory; Nuclear physics; Biological evolution; Germ theory; Industrial revolution; Molecular biology; Information and communication ; Quantum theory; Galactic universe; Medical and health technology.

PS 1d Radioactive isotopes are unstable and undergo spontaneous nuclear reactions, emitting particles and/or wavelike radiation. The decay of any one nucleus cannot be predicted, but a large group of identical nuclei decay at a predictable rate. This predictability can be used to estimate the age of materials that contain radioactive isotopes.

UCP 4 Evolution is a series of changes, some gradual and some sporadic, that accounts for the present form and function of objects, organisms, and natural and designed systems. The general idea of evolution is that the present arises from materials and forms of the past. Although evolution is most commonly associated with the biological theory explaining the process of descent with modification of organisms from common ancestors, evolution also describes changes in the universe.

KEY **SE** = Student Edition **TE** = Teacher Edition **CRF** = Chapter Resource File

Lesson Plan

FOCUS *(5 minutes)*

__ **Overview, TE** Review the objectives listed in the Student Edition. **(GENERAL)**

__ **PowerPoint® Resources** Use the customizable presentation to enhance your lecture.

__ **Bellringer, TE** Students are asked to list the ways they know to determine the exact, or absolute, age of a person or object. **(GENERAL)**

__ **Bellringer Transparency** Use this transparency as students enter the classroom and find their seats.

MOTIVATE *(10 minutes)*

__ **Using the Figure, Erosion Rate, TE** Students use the photo of Niagara Falls to appreciate the size of the Canadian portion of Niagara Falls. **(GENERAL)**

TEACH *(20 minutes)*

__ **Directed Reading, CRF** These worksheets reinforce basic concepts and vocabulary presented in the section. **(BASIC)**

__ **Geography Connection, Niagara Falls, TE** Using a map of North America, students locate Niagara Falls and trace the source and eventual outflow of the falls' water. **(GENERAL)**

__ **Activity, Modeling Varves, TE** Students use light and dark sands and a glass cylinder to build models of varves and answer several questions about varves. **(BASIC)**

__ **Geology Connection, Glacial Lakes, TE** Students research the formation and characteristics of glacial lakes. **(GENERAL)**

__ **Discussion, Measuring Decay, TE** Students discuss why knowing the rate of radioactive decay of different elements is crucial to scientists. **(GENERAL)**

__ **Chemistry Connection, Radioactive Decay of Uranium, TE** Students discuss and answer several questions about radioactive decay. **(ADVANCED)**

__ **Skill Builder, Vocabulary, TE** Students review the word *isotope*. **(ADVANCED)**

__ **Teaching Transparency 42, Radioactive Decay and Half-Life with Worksheet** Use this Teaching Transparency and the accompanying worksheet to support your classroom discussion of radioactive decay and half-life.

__ **Using the Figure, Counting Decay, TE** Students use the figures of isotopes to review half-life. **(BASIC)**

__ **Quick Lab, Radioactive Decay, SE** Students use paper, scissors, and a clock to create a paper isotope. **(GENERAL)**

__ **Datasheet for Quick Lab, Radioactive Decay, CRF** Students use the datasheet to complete the Quick Lab. **(GENERAL)**

__ **Using the Table, Methods for Dating Rocks, TE** Students use the table of radiometric dating methods to review dating rocks. **(GENERAL)**

__ **Environmental Science Connection, Radioactive Waste, TE** Students research the problem of radioactive waste storage and debate the issue of nuclear power. **(ADVANCED)**

KEY **SE** = Student Edition **TE** = Teacher Edition **CRF** = Chapter Resource File

Lesson Plan

CLOSE *(10 minutes)*

__ **Section Review, SE** Students answer end-of-section vocabulary, key ideas, critical thinking, and concept mapping questions. **(GENERAL)**

__ **Reteaching, Half-Life, TE** Students use pennies to conduct experiments about half-life. **(BASIC)**

__ **Quiz, TE** Students answer 2 questions about this section. **(GENERAL)**

__ **Section Quiz, CRF** Use this quiz to assess students' understanding of the section. **(GENERAL) Also in Spanish**

__ **Spanish Section Quiz, Spanish Assessments** Use this quiz to assess students' understanding of the section. **(GENERAL)**

__ **Alternative Assessment, Dating Rock, TE** Students write descriptions of methods they would use to date various rock samples. **(GENERAL)**

KEY **SE** = Student Edition **TE** = Teacher Edition **CRF** = Chapter Resource File

Lesson Plan

Section 3 The Fossil Record

PACING

Regular Schedule:	**with lab(s): 2 days**	**without lab(s): 1 day**
Block Schedule:	**with lab(s): 1 day**	**without lab(s): 0.5 day**

OBJECTIVES

1. **Describe** four ways in which entire organisms can be preserved as fossils.
2. **List** five examples of fossilized traces of organisms.
3. **Describe** how index fossils can be used to determine the age of rocks.

NATIONAL SCIENCE EDUCATION STANDARDS

ES 3b Geologic time can be estimated by observing rock sequences and using fossils to correlate the sequences at various locations. Current methods include using the known decay rates of radioactive isotopes present in rocks to measure the time since the rock was formed.

HNS 3c Occasionally, there are advances in science and technology that have important and long-lasting effects on science and society. Examples of such advances include the following: Copernican revolution; Newtonian mechanics; Relativity; Geologic time scale; Plate tectonics; Atomic theory; Nuclear physics; Biological evolution; Germ theory; Industrial revolution; Molecular biology; Information and communication ; Quantum theory; Galactic universe; Medical and health technology.

LS 3a Species evolve over time. Evolution is the consequence of the interactions of (1) the potential for a species to increase its numbers, (2) the genetic variability of offspring due to mutation and recombination of genes, (3) a finite supply of the resources required for life, and (4) the ensuing selection by the environment of those offspring better able to survive and leave offspring.

LS 3b The great diversity of organisms is the result of more than 3.5 billion years of evolution that has filled every available niche with life forms.

KEY **SE** = Student Edition **TE** = Teacher Edition **CRF** = Chapter Resource File

Lesson Plan

UCP 4 Evolution is a series of changes, some gradual and some sporadic, that accounts for the present form and function of objects, organisms, and natural and designed systems. The general idea of evolution is that the present arises from materials and forms of the past. Although evolution is most commonly associated with the biological theory explaining the process of descent with modification of organisms from common ancestors, evolution also describes changes in the universe.

FOCUS *(5 minutes)*

__ **Overview, TE** Review the objectives listed in the Student Edition. **(GENERAL)**

__ **PowerPoint® Resources** Use the customizable presentation to enhance your lecture.

__ **Bellringer, TE** Have students draw or describe the process by which they think fossils form. **(GENERAL)**

__ **Bellringer Transparency** Use this transparency as students enter the classroom and find their seats.

MOTIVATE *(10 minutes)*

__ **Demonstration, Learning Earth's History, TE** Students look at a fossil and try to determine the history of the fossil. **(GENERAL)**

TEACH *(65 minutes)*

__ **Directed Reading, CRF** These worksheets reinforce basic concepts and vocabulary presented in the section. **(BASIC)**

__ **Discussion, Handprint History, TE** Students imagine that they make cement handprints every year and predict what changes each imprint would show in 1,000 years. **(GENERAL)**

__ **Group Activity, Fossils in Many Forms, TE** In groups, students research one of the five ways fossils form. **(GENERAL)**

__ **Reading Skill Builder, Paired Summarizing, TE** Students read the text in pairs and take turns summarizing what they have read. **(BASIC)**

__ **Activity, Spacing Footprints, TE** Students run and walk, mark and measure the footsteps, and graph the data to show the spacing of their footprints. **(GENERAL)**

__ **Biology Connection, Dinosaurs and Birds, TE** Students research the connection between birds and dinosaurs. **(ADVANCED)**

__ **Making Models Lab, Types of Fossils, SE** Students make models of fossils. **(GENERAL)**

__ **Datasheet for Chapter Lab, Making Models Lab, Types of Fossils, CRF** Students use the datasheet to complete the Chapter Lab. **(GENERAL)**

__ **Mapping Expeditions, Where the Hippos Roam, SE** Students use data from a fossil site, create a map of that site, and draw conclusions about its past environment. **(GENERAL)**

KEY **SE** = Student Edition **TE** = Teacher Edition **CRF** = Chapter Resource File

__ **Inquiry Lab, Got Fossils? CRF** Students identify whether or not various objects are fossils. **(GENERAL)**

__ **Interactive Tutor CD-ROM, Evolution and Geologic Time** Students practice what they have learned using this interactive activity. **(GENERAL)**

CLOSE *(10 minutes)*

__ **Section Review, SE** Students answer end-of-section vocabulary, key ideas, critical thinking, and concept mapping questions. **(GENERAL)**

__ **Reteaching, Listing, TE** Students review the section and list types of fossils, how fossils form, and index fossils. **(BASIC)**

__ **Quiz, TE** Students answer 3 questions about this section. **(GENERAL)**

__ **Section Quiz, CRF** Use this quiz to assess students' understanding of the section. **(GENERAL) Also in Spanish**

__ **Spanish Section Quiz, Spanish Assessments** Use this quiz to assess students' understanding of the section. **(GENERAL)**

__ **Alternative Assessment, Captioned Pictures, TE** Students create booklets about fossils. **(BASIC)**

KEY **SE** = Student Edition **TE** = Teacher Edition **CRF** = Chapter Resource File

End of Chapter Review and Assessment

PACING

Regular Schedule:	**with lab(s): N/A**	**without lab(s): 2 days**
Block Schedule:	**with lab(s): N/A**	**without lab(s): 1 day**

__ **Chapter Review, SE** Students answer end-of-chapter vocabulary, key ideas, critical thinking, math, writing, and interpreting graphics questions. **(GENERAL)**

__ **Alternative Assessment, Modeling Rock Formations, TE** Students use clay, beans, and a clear container to create rock formation models. **(GENERAL)**

__ **Standardized Test Prep, SE** Have students complete the standardized test prep to help them prepare for standardized tests. **(GENERAL)**

__ **Holt PuzzlePro®** Use this software to create crossword puzzles and word searches that make learning vocabulary fun. **(GENERAL)**

__ **Concept Review, CRF** Use this worksheet to assess students' understanding of the main concepts presented in the chapter. **(GENERAL)**

__ **Critical Thinking, CRF** Students use this worksheet to develop and practice critical thinking skills. **(ADVANCED)**

__ **Math Skills, CRF** This worksheet helps reinforce students' understanding of algebraic rearrangements. **(GENERAL)**

__ **Graphing Skills, CRF** This worksheet helps reinforce students' understanding of bar graphs. **(GENERAL)**

__ **Concept Review, Study Guide, The Rock Record** Use this worksheet to assess student's understanding of the main concepts presented in the chapter. **(GENERAL)**

__ **Concept Review, Spanish Study Guide, The Rock Record** This worksheet helps reinforce English-language learners' understanding of the main concepts presented in the chapter. **(GENERAL)**

__ **Chapter Test A, CRF** Assign questions from the appropriate test for chapter assessment. **(GENERAL)**

__ **Chapter Test B, CRF** Assign questions from the appropriate test for chapter assessment. **(ADVANCED)**

__ **Spanish Chapter Test A, Spanish Assessments** Assign questions from the appropriate test for chapter assessment. **(GENERAL)**

__ **Test Generator, One-Stop Planner** Create a customized homework assignment, quiz, or test using the HRW Test Generator program. **(GENERAL)**

__ **Test Item Listing, CRF** Use the Test Item Listing to identify questions for a customized homework, quiz, or test.

__ **PowerPoint® Resources** Use the customizable presentation to help students practice standardized test-taking skills.

KEY **SE** = Student Edition **TE** = Teacher Edition **CRF** = Chapter Resource File

OTHER RESOURCE OPTIONS

__ **HRW Earth Science Video,** Segment 12, "Determining Relative Age" This segment provides an overview of the basic principles geologists use to determine the relative ages of rocks. **(GENERAL)**

__ **HRW Earth Science Video,** Segment 13, "Field Trip—Understanding Rock Layers" In this segment, Dr. Robert Roback takes us on a field trip to a region where igneous, metamorphic, and sedimentary rocks crop out, where he proceeds to reconstruct the area's complex geological history. **(GENERAL)**

__ **Brain Food Video Quiz** Use this video resource to assess student's progress. **(GENERAL)**

KEY **SE** = Student Edition **TE** = Teacher Edition **CRF** = Chapter Resource File

Chapter 9 A View of Earth's Past

Chapter Opener

__ **Pre-Reading Activity, Two-Panel Flip Chart, SE**

__ **Using the Figure, Hypacrosaurus, TE (GENERAL)**

Section 1 Geologic Time

PACING

Regular Schedule:	**with lab(s): 2 days**	**without lab(s): 1 day**
Block Schedule:	**with lab(s): 1 day**	**without lab(s): 0.5 day**

OBJECTIVES

1. **Summarize** how scientists worked together to develop the geologic column.
2. **List** the major divisions of geologic time.

NATIONAL SCIENCE EDUCATION STANDARDS

ES 3b Geologic time can be estimated by observing rock sequences and using fossils to correlate the sequences at various locations. Current methods include using the known decay rates of radioactive isotopes present in rocks to measure the time since the rock was formed.

ES 3d Evidence for one-celled forms of life—the bacteria—extends back more than 3.5 billion years. The evolution of life caused dramatic changes in the composition of the earth's atmosphere, which did not originally contain oxygen.

HNS 3c Occasionally, there are advances in science and technology that have important and long-lasting effects on science and society. Examples of such advances include the following: Copernican revolution; Newtonian mechanics; Relativity; Geologic time scale; Plate tectonics; Atomic theory; Nuclear physics; Biological evolution; Germ theory; Industrial revolution; Molecular biology; Information and communication; Quantum theory; Galactic universe; Medical and health technology

LS 3a Species evolve over time. Evolution is the consequence of the interactions of (1) the potential for a species to increase its numbers, (2) the genetic variability of offspring due to mutation and recombination of genes, (3) a finite supply of the resources required for life, and (4) the ensuing selection by the environment of those offspring better able to survive and leave offspring.

KEY **SE** = Student Edition **TE** = Teacher Edition **CRF** = Chapter Resource File

Lesson Plan

LS 3b The great diversity of organisms is the result of more than 3.5 billion years of evolution that has filled every available niche with life forms.

LS 3c Natural selection and its evolutionary consequences provide a scientific explanation for the fossil record of ancient life forms, as well as for the striking molecular similarities observed among the diverse species of living organisms.

SAI 2c Scientists rely on technology to enhance the gathering and manipulation of data. New techniques and tools provide new evidence to guide inquiry and new methods to gather data, thereby contributing to the advance of science. The accuracy and precision of the data, and therefore the quality of the exploration, depends on the technology used.

FOCUS *(5 minutes)*

__ **Overview, TE** Review the objectives listed in the Student Edition. **(GENERAL)**

__ **PowerPoint® Resources** Use the customizable presentation to enhance your lecture.

__ **Bellringer, TE** Ask students to list what types of changes they think might have taken place on Earth during its more than 4-billion year history. **(GENERAL)**

__ **Bellringer Transparency** Use this transparency as students enter the classroom and find their seats.

MOTIVATE *(10 minutes)*

__ **Using the Figure, Rock Layers, TE** Students use the figures of rock layers to prompt a discussion of geologic time. **(GENERAL)**

TEACH *(65 minutes)*

__ **Directed Reading, CRF** These worksheets reinforce basic concepts and vocabulary presented in the section. **(BASIC)**

__ **Discussion, Dating Rocks, TE** Students discuss and research radiometric dating technology. **(ADVANCED)**

__ **Quick Lab, Geologic Time Scale, SE** Students use paper, a meterstick, and a metric ruler to create a geologic time table. **(GENERAL)**

__ **Datasheet for Quick Lab, Geologic Time Scale, CRF** Students use the datasheet to complete the Quick Lab. **(GENERAL)**

__ **Skill Builder, Vocabulary, TE** Students review the suffix *-zoic*. **(GENERAL)**

__ **Discussion, Making Predictions, TE** Students predict what contemporary objects will be found as fossils a million years from now. **(GENERAL)**

__ **Teaching Transparency 44, The Geologic Time Scale with Worksheet** Use this Teaching Transparency and the accompanying worksheet to support your classroom discussion of the geologic time scale.

KEY **SE** = Student Edition **TE** = Teacher Edition **CRF** = Chapter Resource File

__ **Internet Activity, Extinct Organisms, TE** In groups, students use the internet to find pictures of organisms that were common during one geologic period. **(GENERAL)**

__ **Internet Activity, Extinct Organisms, CRF** This worksheet directs student research for the topic presented in the Teacher Edition. **(GENERAL)**

__ **Skill Practice Lab, History in the Rocks, SE** Students learn how an area's geologic history can be determined by studying its fossils and rock layers. **(GENERAL)**

__ **Datasheet for Chapter Lab, Skill Practice Lab, History in the Rocks, CRF** Students use the datasheet to complete the Chapter Lab. **(GENERAL)**

__ **Group Activity, Dinosaur Map, TE** In groups, students research the distribution and species of dinosaurs that lived in North America, Europe, or Asia. **(GENERAL)**

__ **Science and Technology, CT Scanning Fossils, SE** This article discusses how paleontologists use CT scans to study rare and fragile fossils.

__ **Internet Activity, Imaging Technologies, TE** Students research CT scans and MRI technology and predict how paleontologists could use this technology. **(ADVANCED)**

__ **Internet Activity, Imaging Technologies, CRF** This worksheet directs student research for the topic presented in the Teacher Edition. **(ADVANCED)**

__ **Making Models Lab, Future Earth, CRF** Students model the continents' past movements and predict their future movement. **(GENERAL)**

CLOSE *(10 minutes)*

__ **Section Review, SE** Students answer end-of-section vocabulary, key ideas, critical thinking, and concept mapping questions. **(GENERAL)**

__ **Reteaching, Relative Time, TE** Students create posters showing the relative scale of each geologic eon and era from this chapter. **(BASIC)**

__ **Quiz, TE** Students answer 3 questions about this section. **(GENERAL)**

__ **Section Quiz, CRF** Use this quiz to assess students' understanding of the section. **(GENERAL) Also in Spanish**

__ **Spanish Section Quiz, Spanish Assessments** Use this quiz to assess students' understanding of the section. **(GENERAL)**

__ **Alternative Assessment, Description Essay, TE** Students write essays describing the history of life on Earth. **(ADVANCED)**

KEY **SE** = Student Edition **TE** = Teacher Edition **CRF** = Chapter Resource File

Lesson Plan

Section 2 Precambrian Time and the Paleozoic Era

PACING

Regular Schedule:	**with lab(s): N/A**	**without lab(s): 2 days**
Block Schedule:	**with lab(s): N/A**	**without lab(s): 1 day**

OBJECTIVES

1. **Summarize** how evolution is related to geologic change.
2. **Identify** two characteristics of Precambrian rock.
3. **Identify** one major geologic and two major biological developments during the Paleozoic Era.

NATIONAL SCIENCE EDUCATION STANDARDS

ES 3d Evidence for one-celled forms of life—the bacteria—extends back more than 3.5 billion years. The evolution of life caused dramatic changes in the composition of the earth's atmosphere, which did not originally contain oxygen.

LS 3a Species evolve over time. Evolution is the consequence of the interactions of (1) the potential for a species to increase its numbers, (2) the genetic variability of offspring due to mutation and recombination of genes, (3) a finite supply of the resources required for life, and (4) the ensuing selection by the environment of those offspring better able to survive and leave offspring.

LS 3b The great diversity of organisms is the result of more than 3.5 billion years of evolution that has filled every available niche with life forms.

LS 3c Natural selection and its evolutionary consequences provide a scientific explanation for the fossil record of ancient life forms, as well as for the striking molecular similarities observed among the diverse species of living organisms.

FOCUS *(5 minutes)*

__ **Overview, TE** Review the objectives listed in the Student Edition. **(GENERAL)**

__ **PowerPoint® Resources** Use the customizable presentation to enhance your lecture.

__ **Bellringer, TE** Ask students to speculate how changing conditions on Earth would lead to physical changes in plants and animals. **(GENERAL)**

__ **Bellringer Transparency** Use this transparency as students enter the classroom and find their seats.

KEY **SE** = Student Edition **TE** = Teacher Edition **CRF** = Chapter Resource File

Lesson Plan

MOTIVATE *(10 minutes)*

__ **Using the Figure, Comparing Bones, TE** Students use the figure to describe similarities and differences in the bones of the four animals. **(GENERAL)**

TEACH *(20 minutes)*

__ **Directed Reading, CRF** These worksheets reinforce basic concepts and vocabulary presented in the section. **(BASIC)**

__ **Teaching Transparency 45, Similar Skeletal Structures of Mammals with Worksheet** Use this Teaching Transparency and the accompanying worksheet to support your classroom discussion of similar skeletal structures of mammals.

__ **Teaching Transparency 46, Precambrian Time and the Paleozoic Era with Worksheet** Use this Teaching Transparency and the accompanying worksheet to support your classroom discussion of Precambrian time and the Paleozoic Era.

__ **Biology Connection, Natural Selection, TE** Students research how genes influence natural selection. **(ADVANCED)**

__ **Quick Lab, Chocolate Candy Survival, SE** Students use candy-coated chocolate bits and patterned cloth to see how camouflage protects animals. **(GENERAL)**

__ **Datasheet for Quick Lab, Chocolate Candy Survival, CRF** Students use the datasheet to complete the Quick Lab. **(GENERAL)**

__ **Discussion, Drawing Conclusions, TE** Discuss with students why stromatolites are the most abundant Precambrian fossils. **(GENERAL)**

__ **Internet Activity, Supercontinents, TE** In groups, students research the supercontinents that preceded Pangaea. **(ADVANCED)**

__ **Internet Activity, Supercontinents, CRF** This worksheet directs student research for the topic presented in the Teacher Edition. **(ADVANCED)**

__ **Discussion, Body of Evidence, TE** Discuss with students why some ancient animals are well fossilized and others are not. **(BASIC)**

__ **Using the Figure, Sea to Land, TE** Students use the figures to compare a eurypterid to a modern scorpion. **(GENERAL)**

__ **Graphic Organizer, Spider Map, SE** Use the graphic organizer to help students develop their analytical skills. **(GENERAL)**

__ **Debate, The Current Extinction? TE** In two groups, students research and debate whether or not we are currently undergoing a mass extinction. **(ADVANCED)**

CLOSE *(10 minutes)*

__ **Section Review, SE** Students answer end-of-section vocabulary, key ideas, critical thinking, and concept mapping questions. **(GENERAL)**

__ **Homework, Survivor: The Real Story, TE** Students research blue-green algae and bacteria to discover why these life-forms are so successful at survival. **(ADVANCED)**

KEY **SE** = Student Edition **TE** = Teacher Edition **CRF** = Chapter Resource File

__ **Reteaching, Life in the Paleozoic, TE** Using a long sheet of paper, students illustrate the periods of the Paleozoic Era. **(BASIC)**

__ **Quiz, TE** Students answer 3 questions about this section. **(GENERAL)**

__ **Section Quiz, CRF** Use this quiz to assess students' understanding of the section. **(GENERAL) Also in Spanish**

__ **Spanish Section Quiz, Spanish Assessments** Use this quiz to assess students' understanding of the section. **(GENERAL)**

__ **Alternative Assessment, Timeline, TE** In pairs, students create timelines of the Mesozoic and Cenozoic Eras. **(GENERAL)**

KEY **SE** = Student Edition **TE** = Teacher Edition **CRF** = Chapter Resource File

Lesson Plan

Section 3 The Mesozoic and Cenozoic Eras

PACING

Regular Schedule:	**with lab(s): N/A**	**without lab(s): 2 days**
Block Schedule:	**with lab(s): N/A**	**without lab(s): 1 day**

OBJECTIVES

1. **List** the periods of the Mesozoic and Cenozoic Eras.
2. **Identify** two major geologic and biological developments during the Mesozoic Era.
3. **Identify** two major geologic and biological developments during the Cenozoic Era.

NATIONAL SCIENCE EDUCATION STANDARDS

LS 3c Natural selection and its evolutionary consequences provide a scientific explanation for the fossil record of ancient life forms, as well as for the striking molecular similarities observed among the diverse species of living things.

FOCUS *(5 minutes)*

__ **Overview, TE** Review the objectives listed in the Student Edition. **(GENERAL)**

__ **PowerPoint® Resources** Use the customizable presentation to enhance your lecture.

__ **Bellringer, TE** Ask students to identify movies, TV shows, or books in which dinosaurs or the Mesozoic Era have been portrayed. **(GENERAL)**

__ **Bellringer Transparency** Use this transparency as students enter the classroom and find their seats.

MOTIVATE *(10 minutes)*

__ **Discussion, More Resources? TE** Students discuss why there would be more resources available after a mass extinction. **(GENERAL)**

TEACH *(20 minutes)*

__ **Directed Reading, CRF** These worksheets reinforce basic concepts and vocabulary presented in the section. **(BASIC)**

__ **Teaching Transparency 47, The Mesozoic and Cenozoic Eras with Worksheet** Use this Teaching Transparency and the accompanying worksheet to support your classroom discussion of the Mesozoic and Cenozoic Eras.

__ **Using the Figure, Comparing Characteristics, TE** Students use the figure to compare an Archaeopteryx fossil to modern birds. **(GENERAL)**

__ **Reading Skill Builder, Reading Organizer, TE** Students make a three-column chart to review characteristics of each period of the Mesozoic Era. **(BASIC)**

KEY **SE** = Student Edition **TE** = Teacher Edition **CRF** = Chapter Resource File

__ **Internet Activity, On-line Museums, TE** Students use the internet to find and study museum web sites with dinosaur exhibits. **(GENERAL)**

__ **Internet Activity, On-line Museums, CRF** This worksheet directs student research for the topic presented in the Teacher Edition. **(GENERAL)**

__ **Discussion, Flowering Plants, TE** Students list some flowering plants and animals that depend on them. **(GENERAL)**

__ **Debate, Dinosaur Extinction, TE** Students debate the causes of the mass extinction at the end of the Cretaceous Period. **(ADVANCED)**

__ **Skill Builder, Vocabulary, TE** Students review the meaning of the word *dinosaur*. **(GENERAL)**

__ **Using the Figure, Mammal Characteristics, TE** Students use the figure of a tarsier to describe its characteristics. **(GENERAL)**

__ **Biology Connection, Primitive Mammals, TE** Students research the evolution of *Monotremes* and *Marsupials* to learn how they differ from placental mammals. **(ADVANCED)**

__ **Environmental Science Connection, The Isolation of Antarctica, TE** Students discuss how glaciation has affected Earth, the ocean, and Earth's climate. **(ADVANCED)**

__ **Skill Builder, Math, TE** Students calculate how long mammals existed on Earth before they began to assume a dominant role. **(BASIC)**

__ **Mapping Expeditions, Where the Hippos Roam, SE** Students study data from a fossil site, create a map of that site, and draw conclusions about its past environment. **(GENERAL)**

__ **Group Activity, Dinosaur Map, TE** Students create a dinosaur map. **(GENERAL)**

__ **Maps in Action, Fossil Evidence for Gondwanaland, SE** Students answer questions about a map showing areas where certain fossils have been found to enhance their map-reading skills. **(GENERAL)**

__ **Teaching Transparency 48, Fossil Evidence for Gondwanaland with Worksheet** Use this Teaching Transparency and the accompanying worksheet to help students perform the Maps in Action activity.

__ **Inquiry Lab, Dinosaur Hunt, CRF** Students sort remains from a simulated dinosaur excavation and determine the species and period of the dinosaur. **(GENERAL)**

__ **Interactive Tutor CD-ROM, Evolution and Geologic Time** Students practice what they have learned using this interactive activity. **(GENERAL)**

CLOSE *(10 minutes)*

__ **Section Review, SE** Students answer end-of-section vocabulary, key ideas, critical thinking, and concept mapping questions. **(GENERAL)**

__ **Homework, Horse History The Real Story, TE** Students research the evolution of horses and other members of the genus *Equus*. **(ADVANCED)**

__ **Reteaching, Create a Brochure, TE** In groups, students create travel brochures for the different geologic periods in this section. **(BASIC)**

KEY **SE** = Student Edition **TE** = Teacher Edition **CRF** = Chapter Resource File

Lesson Plan

__ **Quiz, TE** Students answer 2 questions about this section. **(GENERAL)**

__ **Section Quiz, CRF** Use this quiz to assess students' understanding of the section. **(GENERAL) Also in Spanish**

__ **Spanish Section Quiz, Spanish Assessments** Use this quiz to assess students' understanding of the section. **(GENERAL)**

__ **Alternative Assessment, Television Script, TE** Students write television scripts about the Paleozoic Era. **(GENERAL)**

KEY SE = Student Edition **TE** = Teacher Edition **CRF** = Chapter Resource File

Lesson Plan

End of Chapter Review and Assessment

PACING

Regular Schedule:	**with lab(s): N/A**	**without lab(s): 2 days**
Block Schedule:	**with lab(s): N/A**	**without lab(s): 1 day**

__ **Chapter Review, SE** Students answer end-of-chapter vocabulary, key ideas, critical, thinking, math, writing, and interpreting graphics questions. **(GENERAL)**

__ **Alternative Assessment, Science Consultants, TE** Students create a proposal for four theme parks, each representing a geologic era from Precambrian time through the Cenozoic Era. **(ADVANCED)**

__ **Standardized Test Prep, SE** Have students complete the standardized test prep to help them prepare for standardized tests. **(GENERAL)**

__ **Holt PuzzlePro®** Use this software to create crossword puzzles and word searches that make learning vocabulary fun. **(GENERAL)**

__ **Concept Review, CRF** Use this worksheet to assess student's understanding of the main concepts presented in the chapter. **(GENERAL)**

__ **Critical Thinking, CRF** Students use this worksheet to develop and practice critical thinking skills. **(ADVANCED)**

__ **Math Skills, CRF** This worksheet helps reinforce students' understanding of geometry. **(GENERAL)**

__ **Graphing Skills, CRF** This worksheet helps reinforce students' understanding of bar graphs. **(GENERAL)**

__ **Concept Review, Study Guide, A View of Earth's Past** Use this worksheet to assess student's understanding of the main concepts presented in the chapter. **(GENERAL)**

__ **Concept Review, Spanish Study Guide, A View of Earth's Past** This worksheet helps reinforce English-language learners' understanding of the main concepts presented in the chapter. **(GENERAL)**

__ **Chapter Test A, CRF** Assign questions from the appropriate test for chapter assessment. **(GENERAL)**

__ **Chapter Test B, CRF** Assign questions from the appropriate test for chapter assessment. **(ADVANCED)**

__ **Spanish Chapter Test A, Spanish Assessments** Assign questions from the appropriate test for chapter assessment. **(GENERAL)**

__ **Test Generator, One-Stop Planner** Create a customized homework assignment, quiz, or test using the HRW Test Generator program. **(GENERAL)**

__ **Test Item Listing, CRF** Use the Test Item Listing to identify questions for a customized homework, quiz, or test.

__ **PowerPoint® Resources** Use the customizable presentation to help students practice standardized test-taking skills.

KEY **SE** = Student Edition **TE** = Teacher Edition **CRF** = Chapter Resource File

OTHER RESOURCE OPTIONS

__ **Brain Food Video Quiz** Use this video resource to assess student's progress. **(GENERAL)**

KEY **SE** = Student Edition **TE** = Teacher Edition **CRF** = Chapter Resource File

Chapter 10 Plate Tectonics

Chapter Opener

__ **Pre-Reading Activity, Tri-Fold, SE**

__ **Using the Figure, Rifting in Iceland, TE (GENERAL)**

Section 1 Continental Drift

PACING

Regular Schedule:	**with lab(s): 2 days**	**without lab(s): 1 day**
Block Schedule:	**with lab(s): 1 day**	**without lab(s): 0.5 day**

OBJECTIVES

1. **Summarize** Wegener's hypothesis of continental drift.
2. **Describe** the process of sea-floor spreading.
3. **Identify** how paleomagnetism provides support for the idea of sea-floor spreading.
4. **Explain** how sea-floor spreading provides a mechanism for continental drift.

NATIONAL SCIENCE EDUCATION STANDARDS

HNS 3b Usually, changes in science occur as small modifications in extant knowledge. The daily work of science and engineering results in incremental advances in our understanding of the world and our ability to meet human needs and aspirations. Much can be learned about the internal workings of science and the nature of science from study of individual scientists, their daily work, and their efforts to advance scientific knowledge in their area of study.

HNS 3c Occasionally, there are advances in science and technology that have important and long-lasting effects on science and society. Examples of such advances include the following: Copernican revolution; Newtonian mechanics; Relativity; Geologic time scale; Plate tectonics; Atomic theory; Nuclear physics; Biological evolution; Germ theory; Industrial revolution; Molecular biology; Information and communication ; Quantum theory; Galactic universe; Medical and health technology.

SAI 2a Scientists usually inquire about how physical, living, or designed systems function. Conceptual principles and knowledge guide scientific inquiries. Historical and current scientific knowledge influence the design and interpretation of investigations and the evaluation of proposed explanations made by other scientists.

KEY **SE** = Student Edition **TE** = Teacher Edition **CRF** = Chapter Resource File

Lesson Plan

SAI 2e Scientific explanations must adhere to criteria such as: a proposed explanation must be logically consistent; it must abide by the rules of evidence; it must be open to questions and possible modification; and it must be based on historical and current scientific knowledge.

UCP 1 The natural and designed world is complex; it is too large and complicated to investigate and comprehend all at once. Scientists and students learn to define small portions for the convenience of investigation. The units of investigation can be referred to as "systems." A system is an organized group of related objects or components that form a whole. Systems can consist, for example, of organisms, machines, fundamental particles, galaxies, ideas, numbers, transportation, and education. Systems have boundaries, components, resources flow (input and output), and feedback.

FOCUS *(5 minutes)*

__ **Overview, TE** Review the objectives listed in the Student Edition. **(GENERAL)**

__ **PowerPoint® Resources** Use the customizable presentation to enhance your lecture.

__ **Bellringer, TE** Have students propose a hypothesis to explain the similarities between two islands. **(GENERAL)**

__ **Bellringer Transparency** Use this transparency as students enter the classroom and find their seats.

MOTIVATE *(10 minutes)*

__ **Using the Figure, Continental Puzzles, TE** Students use the figure of a world map to see how the continents fit together like puzzle pieces. **(GENERAL)**

TEACH *(65 minutes)*

__ **Directed Reading, CRF** These worksheets reinforce basic concepts and vocabulary presented in the section. **(BASIC)**

__ **Using the Figure, Mountain Ranges and Fossils, TE** Students use the figure of a map to answer questions about mountain ranges and organisms. **(ADVANCED)**

__ **Reading Skill Builder, Paired Summarizing, TE** Students read the text in pairs and take turns summarizing what they have read. **(BASIC)**

__ **Debate, Wegener's Idea, TE** Students research and debate merits and flaws of Wegener's hypothesis. **(ADVANCED)**

__ **Discussion, Making Predictions, TE** Students discuss different mechanisms of continental drift. **(GENERAL)**

__ **Activity, Sea-Floor Sediments, TE** In pairs, students model the depth of sea-floor sediments. **(GENERAL)**

KEY **SE** = Student Edition **TE** = Teacher Edition **CRF** = Chapter Resource File

__ **Graphic Organizer, Chain-of-Events Chart, SE** Use the graphic organizer to help students develop their analytical skills. **(GENERAL)**

__ **Using the Figure, Sea-Floor Formation, TE** Students use the figure of sea-floor spreading to determine the relative age of the blocks in each diagram. **(BASIC)**

__ **Skill Builder, Math, TE** Students calculate how many years it would take for the Atlantic and Pacific Ocean sea floors to spread 1 km. **(ADVANCED)**

__ **Teaching Transparency 49, Sea-Floor Spreading with Worksheet** Use this Teaching Transparency and the accompanying worksheet to support your classroom discussion of sea-floor spreading.

__ **Demonstration, Earth's Magnetic Field, TE** Students observe a demonstration of how iron-bearing minerals in molten rock align with Earth's magnetic field. **(BASIC)**

__ **Discussion, Magnetic Polarity Reversals, TE** Ask students to consider what might happen if Earth's magnetic field were reversed. **(GENERAL)**

__ **Quick Lab, Making Magnets, SE** Students use a bar magnet, an iron nail, and steel paperclips to construct a magnet. **(GENERAL)**

__ **Datasheet for Quick Lab, Making Magnets, CRF** Students use the datasheet to complete the Quick Lab. **(GENERAL)**

__ **Using the Figure, Magnetic Polarity, TE** Students use the figure of the sea floor to discuss magnetic symmetry. **(GENERAL)**

__ **Making Models Lab, Sea-Floor Spreading, SE** Students model the formation of new sea floor at a divergent boundary. **(GENERAL)**

__ **Datasheet for Chapter Lab, Making Models Lab, Sea-Floor Spreading, CRF** Students use the datasheet to complete the Chapter Lab. **(GENERAL)**

__ **Impact on Society, The Mid-Atlantic Ridge, SE** This article discusses the Mid-Atlantic Ridge and how Iceland is affected by sea-floor spreading, seismic, and volcanic activity.

__ **Interactive Tutor CD-ROM, Continental Drift** Students practice what they have learned using this interactive activity. **(GENERAL)**

CLOSE *(10 minutes)*

__ **Section Review, SE** Students answer end-of-section vocabulary, key ideas, critical thinking, and concept mapping questions. **(GENERAL)**

__ **Reteaching, Evidence for Continental Drift, TE** Students build models of the ocean floor. **(BASIC)**

__ **Quiz, TE** Students answer 3 questions about this section. **(GENERAL)**

__ **Section Quiz, CRF** Use this quiz to assess students' understanding of the section. **(GENERAL) Also in Spanish**

__ **Spanish Section Quiz, Spanish Assessments** Use this quiz to assess students' understanding of the section. **(GENERAL)**

__ **Alternative Assessment, Persuasive Essay, TE** Students write essays explaining their stance on sea-floor spreading as a mechanism for continental drift. **(GENERAL)**

KEY **SE** = Student Edition **TE** = Teacher Edition **CRF** = Chapter Resource File

Lesson Plan

Section 2 The Theory of Plate Tectonics

PACING

Regular Schedule:	**with lab(s): N/A**	**without lab(s): 2 days**
Block Schedule:	**with lab(s): N/A**	**without lab(s): 1 day**

OBJECTIVES

1. **Summarize** the theory of plate tectonics.
2. **Identify** and describe the three types of plate boundaries.
3. **List** and describe three causes of plate movement.

NATIONAL SCIENCE EDUCATION STANDARDS

ES 1b The outward transfer of earth's internal heat drives convection circulation in the mantle that propels the plates comprosing earth's surface across the face of the globe.

ES 3c Interactions among the solid earth, the oceans, the atmosphere, and organisms have resulted in the ongoing evolution of the earth system. We can observe some changes such as earthquakes and volcanic eruptions on a human time scale, but many processes such as mountain building and plate movements take place over hundreds of millions of years.

HNS 3c Occasionally, there are advances in science and technology that have important and long-lasting effects on science and society. Examples of such advances include the following: Copernican revolution; Newtonian mechanics; Relativity; Geologic time scale; Plate tectonics; Atomic theory; Nuclear physics; Biological evolution; Germ theory; Industrial revolution; Molecular biology; Information and communication ; Quantum theory; Galactic universe; Medical and health technology.

UCP 3 Although most things are in the process of becoming different—changing—some properties of objects and processes are characterized by constancy, including the speed of light, the charge of an electron, and the total mass plus energy in the universe. Changes might occur, for example, in properties of materials, position of objects, motion, and form and function of systems. Interactions within and among systems result in change. Changes vary in rate, scale, and pattern, including trends and cycles.

FOCUS *(5 minutes)*

__ **Overview, TE** Review the objectives listed in the Student Edition. **(GENERAL)**

__ **PowerPoint® Resources** Use the customizable presentation to enhance your lecture.

KEY **SE** = Student Edition **TE** = Teacher Edition **CRF** = Chapter Resource File

Lesson Plan

__ **Bellringer, TE** Ask students how many continents exist on Earth's surface, and have them list the continents. **(GENERAL)**

__ **Bellringer Transparency** Use this transparency as students enter the classroom and find their seats.

MOTIVATE *(10 minutes)*

__ **Activity, Jigsaw Puzzle, TE** Students compare Earth's surface to a jigsaw puzzle. **(GENERAL)**

TEACH *(20 minutes)*

__ **Directed Reading, CRF** These worksheets reinforce basic concepts and vocabulary presented in the section. **(BASIC)**

__ **Skill Builder, Vocabulary, TE** Students review the origins of the word *tectonics.* **(BASIC)**

__ **Internet Activity, Earthquakes, TE** Students research areas of Earth's surface that have the most earthquakes. **(BASIC)**

__ **Internet Activity, Earthquakes, CRF** This worksheet directs student research for the topic presented in the Teacher Edition. **(BASIC)**

__ **Physics Connection, Earthquakes, TE** Students review what happens during an earthquake.

__ **Teaching Transparency 50, Tectonic Plate Boundaries with Worksheet** Use this Teaching Transparency and the accompanying worksheet to support your classroom discussion of tectonic plate boundaries.

__ **Discussion, Crustal Chemistry, TE** Students review the composition of oceanic and continental crust to understand convergent boundaries. **(BASIC)**

__ **Skill Builder, Vocabulary, TE** Students review the root of the word *divergent.* **(BASIC)**

__ **Teaching Transparency 51, Types of Plate Boundaries with Worksheet** Use this Teaching Transparency and the accompanying worksheet to support your classroom discussion of types of plate boundaries.

__ **Skill Builder, Writing, TE** Students research and write an essay about a specific plate boundary. **(GENERAL)**

__ **Using the Figure, Mantle Convection, TE** Students use the figure to analyze how the areas where lithosphere dives into the mantle and where colder mantle material sinks are related. **(GENERAL)**

__ **Quick Lab, Tectonic Plate Boundaries, SE** Students conduct an experiment to study tectonic plate boundaries. **(GENERAL)**

__ **Datasheet for Quick Lab, Tectonic Plate Boundaries, CRF** Students use the datasheet to complete the Quick Lab. **(GENERAL)**

__ **Teaching Transparency 52, Ridge Push and Slab Pull with Worksheet** Use this Teaching Transparency and the accompanying worksheet to support your classroom discussion of ridge push and slab pull.

KEY **SE** = Student Edition **TE** = Teacher Edition **CRF** = Chapter Resource File

Lesson Plan

__ **Maps in Action, Locations of Earthquakes in South America 2002–2003, SE** Students enhance their map-reading skills by answering questions about a map showing the locations and depths of earthquakes in South America. **(GENERAL)**

__ **Teaching Transparency 54, Locations of Earthquakes in South America, 2002–2003 with Worksheet** Use this Teaching Transparency and the accompanying worksheet to help students perform the Maps in Action activity.

__ **Internet Activity, The Heimaey Eruption, TE** Students research the Heimaey Eruption in Iceland and the attempts made to save the town. **(BASIC)**

__ **Internet Activity, The Heimaey Eruption, CRF** This worksheet directs student research for the topic presented in the Teacher Edition. **(BASIC)**

__ **Group Activity, 3-D Model, TE** Students make three-dimensional models of the South American plate boundary. **(ADVANCED)**

__ **Mapping Expeditions, A Case of the Tennessee Shakes, SE** Students study maps of seismic activity in the world and in Tennessee and hypothesize about causes of earthquakes that occur away from fault lines. **(GENERAL)**

__ **Inquiry Lab, Where Do Earthquakes Happen? CRF** Students collect and organize earthquake data to study how their locations relate to tectonic plate boundaries. **(GENERAL)**

__ **Making Models Lab, Eggshell Tectonics, CRF** Students use hard-boiled eggs to illustrate and examine tectonic plate movement. **(GENERAL)**

__ **Interactive Tutor CD-ROM, Tectonic Plates** Students practice what they have learned using this interactive activity. **(GENERAL)**

CLOSE *(10 minutes)*

__ **Section Review, SE** Students answer end-of-section vocabulary, key ideas, critical thinking, and concept mapping questions. **(GENERAL)**

__ **Reteaching, Tectonic Processes, TE** Using clay, students build models of two tectonic plates. **(BASIC)**

__ **Quiz, TE** Students answer 3 questions about this section. **(GENERAL)**

__ **Section Quiz, CRF** Use this quiz to assess students' understanding of the section. **(GENERAL) Also in Spanish**

__ **Spanish Section Quiz, Spanish Assessments** Use this quiz to assess students' understanding of the section. **(GENERAL)**

__ **Alternative Assessment, Tectonic Art, TE** Students draw examples of the three types of plate boundaries and label all of the components. **(BASIC)**

KEY SE = Student Edition **TE** = Teacher Edition **CRF** = Chapter Resource File

Lesson Plan

Section 3 The Changing Continents

PACING

Regular Schedule:	**with lab(s): N/A**	**without lab(s): 2 days**
Block Schedule:	**with lab(s): N/A**	**without lab(s): 1 day**

OBJECTIVES

1. **Identify** how movements of tectonic plates change Earth's surface.
2. **Summarize** how movements of tectonic plates have influenced climates and life on Earth.
3. **Describe** the supercontinent cycle.

NATIONAL SCIENCE EDUCATION STANDARDS

ES 3c Interactions among the solid earth, the oceans, the atmosphere, and organisms have resulted in the ongoing evolution of the earth system. We can observe some changes such as earthquakes and volcanic eruptions on a human time scale, but many processes such as mountain building and plate movements take place over hundreds of millions of years.

UCP 2 Evidence consists of observations and data on which to base scientific explanations. Using evidence to understand interactions allows individuals to predict changes in natural and designed systems. Models are tentative schemes or structures that correspond to real objects, events, or classes of events, and that have explanatory power. Models help scientists and engineers understand how things work. Models take many forms, including physical objects, plans, mental constructs, mathematical equations, and computer simulations.

UCP 3 Although most things are in the process of becoming different—changing—some properties of objects and processes are characterized by constancy, including the speed of light, the charge of an electron, and the total mass plus energy in the universe. Changes might occur, for example, in properties of materials, position of objects, motion, and form and function of systems. Interactions within and among systems result in change. Changes vary in rate, scale, and pattern, including trends and cycles.

FOCUS *(5 minutes)*

__ **Overview, TE** Review the objectives listed in the Student Edition. **(GENERAL)**

__ **PowerPoint® Resources** Use the customizable presentation to enhance your lecture.

KEY SE = Student Edition **TE** = Teacher Edition **CRF** = Chapter Resource File

__ **Bellringer, TE** Ask students what the United States might be like if it were centered over the equator or centered over the North Pole. **(GENERAL)**

__ **Bellringer Transparency** Use this transparency as students enter the classroom and find their seats.

MOTIVATE *(10 minutes)*

__ **Activity, Modeling Rifting, TE** Students use clay or putty to demonstrate what happens when continents rift. **(GENERAL)**

TEACH *(20 minutes)*

__ **Directed Reading, CRF** These worksheets reinforce basic concepts and vocabulary presented in the section. **(BASIC)**

__ **Demonstration, Modeling Accretion, TE** Students use colored clay to demonstrate accretion. **(GENERAL)**

__ **Using the Figure, Accretion, TE** Students use the figure of accreting terranes to draw what the continent will look like after the remaining oceanic crust has subducted. **(GENERAL)**

__ **Skill Builder, Vocabulary, TE** Students review the origin of the word *accrete*. **(ADVANCED)**

__ **Group Activity, Responses to Climate Change, TE** In groups, students discuss and explain how their group adapts to climate change. **(ADVANCED)**

__ **Reading Skill Builder, Reading Organizer, TE** Students are encouraged to take notes while reading this section. **(BASIC)**

__ **Teaching Transparency 53, The Supercontinent Cycle with Worksheet** Use this Teaching Transparency and the accompanying worksheet to support your classroom discussion of the supercontinent cycle.

__ **Internet Activity, The Paleomap Project, TE** Students research the Paleomap Project and present their findings. **(BASIC)**

__ **Internet Activity, The Paleomap Project, CRF** This worksheet directs student research for the topic presented in the Teacher Edition. **(BASIC)**

CLOSE *(10 minutes)*

__ **Section Review, SE** Students answer end-of-section vocabulary, key ideas, critical thinking, and concept mapping questions. **(GENERAL)**

__ **Reteaching, Peer Reviewing, TE** In small groups, students compose five questions about the section and quiz each other. **(BASIC)**

__ **Quiz, TE** Students answer 3 questions about this section. **(GENERAL)**

__ **Section Quiz, CRF** Use this quiz to assess students' understanding of the section. **(GENERAL) Also in Spanish**

__ **Spanish Section Quiz, Spanish Assessments** Use this quiz to assess students' understanding of the section. **(GENERAL)**

KEY **SE** = Student Edition **TE** = Teacher Edition **CRF** = Chapter Resource File

__ **Alternative Assessment, Process Models, TE** Students create models illustrating either rifting or accretion. **(ADVANCED)**

KEY **SE** = Student Edition **TE** = Teacher Edition **CRF** = Chapter Resource File

End of Chapter Review and Assessment

PACING

Regular Schedule:	**with lab(s): N/A**	**without lab(s): 2 days**
Block Schedule:	**with lab(s): N/A**	**without lab(s): 1 day**

__ **Chapter Review, SE** Students answer end-of-chapter vocabulary, key ideas, critical thinking, math, writing, and interpreting graphics questions. **(GENERAL)**

__ **Alternative Assessment, Comparing Scientific Ideas, TE** Students create a table that compares Wegener's hypothesis of continental drift with the modern theory of plate tectonics. **(ADVANCED)**

__ **Standardized Test Prep, SE** Have students complete the standardized test prep to help them prepare for standardized tests. **(GENERAL)**

__ **Holt PuzzlePro®** Use this software to create crossword puzzles and word searches that make learning vocabulary fun. **(GENERAL)**

__ **Concept Review, CRF** Use this worksheet to assess student's understanding of the main concepts presented in the chapter. **(GENERAL)**

__ **Critical Thinking, CRF** Students use this worksheet to develop and practice critical thinking skills. **(ADVANCED)**

__ **Math Skills, CRF** This worksheet helps reinforce students' understanding of scientific notation. **(GENERAL)**

__ **Graphing Skills, CRF** This worksheet helps reinforce students' understanding of pie graphs. **(GENERAL)**

__ **Concept Review, Study Guide, Plate Tectonics** Use this worksheet to assess student's understanding of the main concepts presented in the chapter. **(GENERAL)**

__ **Concept Review, Spanish Study Guide, Plate Tectonics** This worksheet helps reinforce English-language learners' understanding of the main concepts presented in the chapter. **(GENERAL)**

__ **Chapter Test A, CRF** Assign questions from the appropriate test for chapter assessment. **(GENERAL)**

__ **Chapter Test B, CRF** Assign questions from the appropriate test for chapter assessment. **(ADVANCED)**

__ **Spanish Chapter Test A, Spanish Assessments** Assign questions from the appropriate test for chapter assessment. **(GENERAL)**

__ **Test Generator, One-Stop Planner** Create a customized homework assignment, quiz, or test using the HRW Test Generator program. **(GENERAL)**

__ **Test Item Listing, CRF** Use the Test Item Listing to identify questions for a customized homework, quiz, or test.

__ **PowerPoint® Resources** Use the customizable presentation to help students practice standardized test-taking skills.

KEY **SE** = Student Edition **TE** = Teacher Edition **CRF** = Chapter Resource File

OTHER RESOURCE OPTIONS

__ **HRW Earth Science Video,** Segment 3, "Plate Tectonics" This segment shows the different layers of Earth, and explains the density difference that causes the cool, rigid lithosphere that is broken up into tectonic plates to float on the hotter, fluid asthenosphere. **(GENERAL)**

__ **NOVA Video, "Earthquake"** This program shows how today's advanced technology helps geologists predict earthquakes. **(GENERAL)**

__ **Brain Food Video Quiz** Use this video resource to assess student's progress. **(GENERAL)**

KEY **SE** = Student Edition **TE** = Teacher Edition **CRF** = Chapter Resource File

Chapter 11 Deformation of the Crust

Chapter Opener

__ **Pre-Reading Activity, Key-Term Fold, SE**

__ **Using the Figure, TE (GENERAL)**

Section 1 How Rock Deforms

PACING

Regular Schedule:	**with lab(s): N/A**	**without lab(s): 2 days**
Block Schedule:	**with lab(s): N/A**	**without lab(s): 1 day**

OBJECTIVES

1. **Summarize** the principle of isostasy.
2. **Identify** the three main types of stress.
3. **Compare** folds and faults.

NATIONAL SCIENCE EDUCATION STANDARDS

HNS 2a: Science distinguishes itself from other ways of knowing and from other bodies of knowledge through the use of empirical standards, logical arguments, and skepticism, as scientists strive for the best possible explanations about the natural world.

UCP 3: Although most things are in the process of becoming different—changing—some properties of objects and processes are characterized by constancy, including the speed of light, the charge of an electron, and the total mass plus energy in the universe. Changes might occur, for example, in properties of materials, position of objects, motion, and form and function of systems. Interactions within and among systems result in change. Changes vary in rate, scale, and pattern, including trends and cycles.

UCP 4: Evolution is a series of changes, some gradual and some sporadic, that accounts for the present form and function of objects, organisms, and natural and designed systems. The general idea of evolution is that the present arises from materials and forms of the past. Although evolution is most commonly associated with the biological theory explaining the process of descent with modification of organisms from common ancestors, evolution also describes changes in the universe.

KEY **SE** = Student Edition **TE** = Teacher Edition **CRF** = Chapter Resource File

Lesson Plan

FOCUS *(5 minutes)*

__ **Overview, TE** Review the objectives listed in the Student Edition. **(GENERAL)**

__ **PowerPoint® Resources** Use the customizable presentation to enhance your lecture.

__ **Bellringer, TE** Have students predict what would happen if they placed a bowling ball on a soft pillow. **(GENERAL)**

__ **Bellringer Transparency** Use this transparency as students enter the classroom and find their seats.

MOTIVATE *(10 minutes)*

__ **Demonstration, Deformation of Earth's Crust, TE** Use a clear container, mud, and a brick to demonstrate isostatic adjustment. **(GENERAL)**

TEACH *(20 minutes)*

__ **Directed Reading, CRF** These worksheets reinforce basic concepts and vocabulary presented in the section. **(BASIC)**

__ **Teaching Transparency 55, Isostatic Adjustment with Worksheet** Use this Teaching Transparency and the accompanying worksheet to support your classroom discussion of isostatic adjustment.

__ **Quick Lab, Modeling Isostasy, SE** Students conduct an experiment to better understand isostasy. **(GENERAL)**

__ **Datasheet for Quick Lab, Modeling Isostasy, CRF** Students use the datasheet to complete the Quick Lab. **(GENERAL)**

__ **Physics Connection, Archimedes' Principle, TE** Students research how Archimedes' principle is used to design ships and balloons. **(ADVANCED)**

__ **Demonstration, Types of Stress, TE** Use a block of clay to demonstrate different kinds of stress. **(BASIC)**

__ **Teaching Transparency 56, Types of Stress with Worksheet** Use this Teaching Transparency and the accompanying worksheet to support your classroom discussion of different types of stress.

__ **Quick Lab, Modeling Stress and Strain, SE** Students use gloves and warm and frozen putty to model stress and strain. **(GENERAL)**

__ **Datasheet for Quick Lab, CRF** Students use the datasheet to complete the Quick Lab. **(GENERAL)**

__ **Skill Builder, Vocabulary, TE** Students review the definitions of *stress* and *strain*. **(BASIC)**

__ **Physics Connection, Elastic Modulus, TE** Students research forms of elastic modulus and relate them to kinds of stress that rock undergoes. **(ADVANCED)**

__ **Using the Figure, Identifying Limbs and Hinges, TE** Students use the figure of a rock to answer questions about hinges and limbs. **(GENERAL)**

KEY **SE** = Student Edition **TE** = Teacher Edition **CRF** = Chapter Resource File

__ **Activity, Folded Art, TE** Students create drawings or sculptures representing symmetric and asymmetric folds. **(ADVANCED)**

__ **Teaching Transparency 57, Folds with Worksheet** Use this Teaching Transparency and the accompanying worksheet to support your classroom discussion of folds.

__ **Reading Skill Builder, Reading Organizer, TE** Students create tables, then record information about folds while reading the page. **(BASIC)**

__ **Teaching Transparency 58, Faults with Worksheet** Use this Teaching Transparency and the accompanying worksheet to support your classroom discussion of faults.

__ **Group Activity, It's Your Fault, TE** Students work in groups to research major fault lines on a continent. **(ADVANCED)**

__ **Maps in Action, Shear Strain in New Zealand, SE** Students answer questions about a map showing the plate boundary zone of New Zealand to enhance their map-reading skills. **(GENERAL)**

__ **Teaching Transparency 61, Shear Strain in New Zealand with Worksheet** Use this Teaching Transparency and the accompanying worksheet to help students perform the Maps in Action activity.

__ **Activity, Geophysics of New Zealand** Students research New Zealand's geology. **(GENERAL)**

__ **Inquiry Lab, Rock Deformation, CRF** Students use weights and a pulley system to examine how changes in stress deform different materials. **(GENERAL)**

__ **Skills Practice Lab, Hooke's Law, CRF** Students use Hooke's Law to observe how a spring's length is affected by applying masses of varying weights. **(GENERAL)**

CLOSE *(10 minutes)*

__ **Section Review, SE** Students answer end-of-section vocabulary, key ideas, critical thinking, and concept mapping questions. **(GENERAL)**

__ **Homework, Locating Types of Folds, TE** Students research areas of the United States where each of the three types of folds are common. **(BASIC)**

__ **Reteaching, Deformation Flashcards, TE** Create flashcards to review stress, strain, folds, and faults with students. **(BASIC)**

__ **Quiz, TE** Students answer 3 questions about this section. **(GENERAL)**

__ **Section Quiz, CRF** Use this quiz to assess students' understanding of the section. **(GENERAL) Also in Spanish**

__ **Spanish Section Quiz, Spanish Assessments** Use this quiz to assess students' understanding of the section. **(GENERAL)**

__ **Alternative Assessment, Modeling Isostasy, TE** Students create models demonstrating what changes occur during an isostatic adjustment. **(ADVANCED)**

KEY **SE** = Student Edition **TE** = Teacher Edition **CRF** = Chapter Resource File

Section 2 How Mountains Form

PACING

Regular Schedule:	**with lab(s): 2 days**	**without lab(s): 1 day**
Block Schedule:	**with lab(s): 1 day**	**without lab(s): 0.5 day**

OBJECTIVES

1. **Identify** the types of plate collisions that form mountains.
2. **Identify** four types of mountains.
3. **Compare** how folded and fault-block mountains form.

NATIONAL SCIENCE EDUCATION STANDARDS

ES 1b: The outward transfer of earth's internal heat drives convection circulation in the mantle that propels the plates comprosing earth's surface across the face of the globe.

ES 3c: Interactions among the solid earth, the oceans, the atmosphere, and organisms have resulted in the ongoing evolution of the earth system. We can observe some changes such as earthquakes and volcanic eruptions on a human time scale, but many processes such as mountain building and plate movements take place over hundreds of millions of years.

HNS 2a: Science distinguishes itself from other ways of knowing and from other bodies of knowledge through the use of empirical standards, logical arguments, and skepticism, as scientists strive for the best possible explanations about the natural world.

UCP 3: Although most things are in the process of becoming different—changing—some properties of objects and processes are characterized by constancy, including the speed of light, the charge of an electron, and the total mass plus energy in the universe. Changes might occur, for example, in properties of materials, position of objects, motion, and form and function of systems. Interactions within and among systems result in change. Changes vary in rate, scale, and pattern, including trends and cycles.

UCP 4: Evolution is a series of changes, some gradual and some sporadic, that accounts for the present form and function of objects, organisms, and natural and designed systems. The general idea of evolution is that the present arises from materials and forms of the past. Although evolution is most commonly associated with the biological theory explaining the process of descent with modification of organisms from common ancestors, evolution also describes changes in the universe.

KEY **SE** = Student Edition **TE** = Teacher Edition **CRF** = Chapter Resource File

Lesson Plan

FOCUS *(5 minutes)*

__ **Overview, TE** Review the objectives listed in the Student Edition. **(GENERAL)**

__ **PowerPoint® Resources** Use the customizable presentation to enhance your lecture.

__ **Bellringer, TE** Have students list as many mountain ranges as they can and name the continents on which the ranges are located. **(GENERAL)**

__ **Bellringer Transparency** Use this transparency as students enter the classroom and find their seats.

MOTIVATE *(10 minutes)*

__ **Discussion, Mountain Features, TE** Discuss features of mountains with students. **(GENERAL)**

TEACH *(65 minutes)*

__ **Directed Reading, CRF** These worksheets reinforce basic concepts and vocabulary presented in the section. **(BASIC)**

__ **Graphic Organizer, Cause-and-Effect Map, SE** Use the graphic organizer to help students develop their analytical skills. **(GENERAL)**

__ **Using the Figure, Lithospheric Collisions and Mountain Types, TE** Students answer questions about the formation of mountain ranges. **(BASIC)**

__ **Teaching Transparency 59, How Mountains Form with Worksheet** Use this Teaching Transparency and the accompanying worksheet to support your classroom discussion of how mountains form.

__ **Skill Builder, Writing, TE** Students read mountain-inspired writings and write a book report. **(GENERAL)**

__ **Group Activity, Plates and Mountain Formation, TE** Students work in groups to research the processes that formed a particular mountain range. **(GENERAL)**

__ **Teaching Transparency 60, Types of Mountains in the United States with Worksheet** Use this Teaching Transparency and the accompanying worksheet to support your classroom discussion of types of mountains in the United States.

__ **Internet Activity, "Dead" Grabens, TE** Students use the internet to research similarities and differences between Death Valley and the Dead Sea. **(ADVANCED)**

__ **Internet Activity, "Dead" Grabens, CRF** This worksheet directs student research for the topic presented in the Teacher Edition. **(ADVANCED)**

__ **Reading Skill Builder, Discussion, TE** Students discuss why the Appalachians are not a high mountain range. **(BASIC)**

__ **Making Models Lab, Continental Collisions, SE** Students build models showing how the Himalayas were formed from colliding tectonic plates. **(GENERAL)**

__ **Datasheet for Chapter Lab, Making Models Lab, Continental Collisions, CRF** Students use the datasheet to complete the Chapter Lab. **(GENERAL)**

KEY **SE** = Student Edition **TE** = Teacher Edition **CRF** = Chapter Resource File

__ **Impact on Society, The Disappearing Mediterranean, SE** The Mediterranean Sea is slowly disappearing due to subduction of the African and Eurasian plates.

__ **Activity, The Mediterranean of the Past, TE** Students research the geologic history of the Mediterranean Sea. **(ADVANCED)**

CLOSE *(10 minutes)*

__ **Section Review, SE** Students answer end-of-section vocabulary, key ideas, critical thinking, and concept mapping questions. **(GENERAL)**

__ **Homework, Valleys, Plateaus, and Grabens, TE** Students research some geologic features and present their findings. **(GENERAL)**

__ **Reteaching, Mountain Formation Processes, TE** Students review how the four main types of mountains are formed. **(BASIC)**

__ **Quiz, TE** Students answer 3 questions about this section. **(GENERAL)**

__ **Section Quiz, CRF** Use this quiz to assess students' understanding of the section. **(GENERAL) Also in Spanish**

__ **Spanish Section Quiz, Spanish Assessments** Use this quiz to assess students' understanding of the section. **(GENERAL)**

__ **Alternative Assessment, Modeling Plate Interactions, TE** Students create models showing how the interactions between tectonic plates cause different types of mountains to form. **(ADVANCED)**

KEY **SE** = Student Edition **TE** = Teacher Edition **CRF** = Chapter Resource File

End of Chapter Review and Assessment

PACING

Regular Schedule:	**with lab(s): N/A**	**without lab(s): 2 days**
Block Schedule:	**with lab(s): N/A**	**without lab(s): 1 day**

__ **Chapter Review, SE** Students answer end-of-chapter vocabulary, key ideas, critical thinking, math, writing, and interpreting graphics questions. **(GENERAL)**

__ **Alternative Assessment, Poster Project, TE** Students create a poster that shows the processes that rocks undergo that produce different types of mountains. **(GENERAL)**

__ **Standardized Test Prep, SE** Have students complete the standardized test prep to help them prepare for standardized tests. **(GENERAL)**

__ **Holt PuzzlePro®** Use this software to create crossword puzzles and word searches that make learning vocabulary fun. **(GENERAL)**

__ **Concept Review, CRF** Use this worksheet to assess student's understanding of the main concepts presented in the chapter. **(GENERAL)**

__ **Critical Thinking, CRF** Students use this worksheet to develop and practice critical thinking skills. **(ADVANCED)**

__ **Math Skills, CRF** This worksheet helps reinforce students' understanding of algebraic rearrangements. **(GENERAL)**

__ **Graphing Skills, CRF** This worksheet helps reinforce students' understanding of bar graphs. **(GENERAL)**

__ **Concept Review, Study Guide, Deformation of the Crust** Use this worksheet to assess student's understanding of the main concepts presented in the chapter. **(GENERAL)**

__ **Concept Review, Spanish Study Guide, Deformation of the Crust** This worksheet helps reinforce English-language learners' understanding of the main concepts presented in the chapter. **(GENERAL)**

__ **Chapter Test A, CRF** Assign questions from the appropriate test for chapter assessment. **(GENERAL)**

__ **Chapter Test B, CRF** Assign questions from the appropriate test for chapter assessment. **(ADVANCED)**

__ **Spanish Chapter Test A, Spanish Assessments** Assign questions from the appropriate test for chapter assessment. **(GENERAL)**

__ **Test Generator, One-Stop Planner** Create a customized homework assignment, quiz, or test using the HRW Test Generator program. **(GENERAL)**

__ **Test Item Listing, CRF** Use the Test Item Listing to identify questions for a customized homework, quiz, or test.

__ **PowerPoint® Resources** Use the customizable presentation to help students practice standardized test-taking skills.

KEY SE = Student Edition **TE** = Teacher Edition **CRF** = Chapter Resource File

OTHER RESOURCE OPTIONS

__ **HRW Earth Science Video, Segment 4, "Deformation of the Crust"** This segment examines the different stresses produced by plate movements in the Earth's crust and the specific stresses that occur different plate boundaries. **(GENERAL)**

__ **Brain Food Video Quiz** Use this video resource to assess student's progress. **(GENERAL)**

KEY **SE** = Student Edition **TE** = Teacher Edition **CRF** = Chapter Resource File

Chapter 12 Earthquakes

Chapter Opener

__ **Pre-Reading Activity, Pyramid, SE**

__ **Using the Figure, The Kobe Quake, TE (GENERAL)**

Section 1 How and Where Earthquakes Happen

PACING

Regular Schedule:	**with lab(s): N/A**	**without lab(s): 2 days**
Block Schedule:	**with lab(s): N/A**	**without lab(s): 1 day**

OBJECTIVES

1. **Describe** elastic rebound.
2. **Compare** body waves and surface waves.
3. **Explain** how the structure of Earth's interior affects seismic waves.
4. **Explain** why earthquakes generally occur at plate boundaries.

NATIONAL SCIENCE EDUCATION STANDARDS

PS 6: Interactions of energy and matter.

PS 6a: Waves, including sound and seismic waves, waves on water, and light waves, have energy and can transfer energy when they interact with matter.

SPSP 5c: Some hazards, such as earthquakes, volcanic eruptions, and severe weather, are rapid and spectacular. But there are slow and progressive changes that also result in problems for individuals and societies. For example, change in stream channel position, erosion of bridge foundations, sedimentation in lakes and harbors, coastal erosions, and continuing erosion and wasting of soil and landscapes can all negatively affect society.

FOCUS *(5 minutes)*

__ **Overview, TE** Review the objectives listed in the Student Edition. **(GENERAL)**

__ **PowerPoint® Resources** Use the customizable presentation to enhance your lecture.

__ **Bellringer, TE** Students describe what happens when a rock is thrown into a pond and how this is similar to an earthquake. **(GENERAL)**

__ **Bellringer Transparency** Use this transparency as students enter the classroom and find their seats.

KEY **SE** = Student Edition **TE** = Teacher Edition **CRF** = Chapter Resource File

Lesson Plan

MOTIVATE *(10 minutes)*

__ **Demonstration, Elastic Rebound, TE** Students observe how bending stores and releases energy. **(GENERAL)**

TEACH *(20 minutes)*

__ **Directed Reading, CRF** These worksheets reinforce basic concepts and vocabulary presented in the section. **(BASIC)**

__ **Group Activity, Modeling Locked Faults, TE** Students use blocks and sandpaper to simulate what happens during an earthquake. **(GENERAL)**

__ **Skill Builder, Vocabulary, TE** Students discuss the origins of the prefixes *epi-* and *hypo-*. **(GENERAL)**

__ **Internet Activity, Distribution Patterns, TE** Students research the distribution of earthquake foci and identify seismically active areas in North America. **(GENERAL)**

__ **Internet Activity, Distribution Patterns, CRF** This worksheet directs student research for the topic presented in the Teacher Edition. **(GENERAL)**

__ **Teaching Transparency 62, Anatomy of an Earthquake with Worksheet** Use this Teaching Transparency and the accompanying worksheet to support your classroom discussion of the anatomy of an earthquake.

__ **Technology Connection, Seismic Profiling, TE** Students create models or diagrams showing how seismic profiling works. **(GENERAL)**

__ **Discussion, Surface Waves, TE** Students explain why surface waves are so damaging. **(BASIC)**

__ **History Connection, Moho Discontinuity, TE** Students research the Croatian scientist Andrija Mohorovi˘ci'c. **(GENERAL)**

__ **Graphic Organizer, Spider Map, SE** Use the graphic organizer to help students develop their analytical skills. **(GENERAL)**

__ **Teaching Transparency 63, Seismic Waves and Earth's Interior with Worksheet** Use this Teaching Transparency and the accompanying worksheet to support your classroom discussion of seismic waves and Earth's interior.

__ **Teaching Transparency 64, Earthquakes and Tectonic Plate Boundaries with Worksheet** Use this Teaching Transparency and the accompanying worksheet to support your classroom discussion of earthquakes and tectonic plate boundaries.

__ **Mapping Expeditions, A Case of the Tennessee Shakes, SE** Students study maps of seismic activity in the world and in Tennessee and hypothesize about causes of earthquakes that occur away from fault lines. **(GENERAL)**

__ **Inquiry Lab, Simulating Earthquakes, CRF** Students test how changes in friction affect plate slip and elastic rebound by modeling how earthquakes originate. **(GENERAL)**

CLOSE *(10 minutes)*

KEY **SE** = Student Edition **TE** = Teacher Edition **CRF** = Chapter Resource File

Lesson Plan

__ **Section Review, SE** Students answer end-of-section vocabulary, key ideas, critical thinking, and concept mapping questions. **(GENERAL)**

__ **Reteaching, Earthquake Maps, TE** Students create flipbooks illustrating elastic rebound. **(BASIC)**

__ **Quiz, TE** Students answer 2 questions about this section. **(GENERAL)**

__ **Section Quiz, CRF** Use this quiz to assess students' understanding of the section. **(GENERAL) Also in Spanish**

__ **Spanish Section Quiz, Spanish Assessments** Use this quiz to assess students' understanding of the section. **(GENERAL)**

__ **Alternative Assessment, Earthquake Patterns, TE** Students identify earthquake patterns from a map showing earthquake activity in the U.S. **(GENERAL)**

KEY **SE** = Student Edition **TE** = Teacher Edition **CRF** = Chapter Resource File

Section 2 Studying Earthquakes

PACING

Regular Schedule:	**with lab(s): 2 days**	**without lab(s): 1 day**
Block Schedule:	**with lab(s): 1 day**	**without lab(s): 0.5 day**

OBJECTIVES

1. **Describe** the instrument used to measure and record earthquakes.
2. **Summarize** the method scientists use to locate an epicenter.
3. **Describe** the scales used to measure the magnitude and intensity of earthquakes.

NATIONAL SCIENCE EDUCATION STANDARDS

PS 6: Interactions of energy and matter

PS 6a: Waves, including sound and seismic waves, waves on water, and light waves, have energy and can transfer energy when they interact with matter.

SPSP 5c: Some hazards, such as earthquakes, volcanic eruptions, and severe weather, are rapid and spectacular. But there are slow and progressive changes that also result in problems for individuals and societies. For example, change in stream channel position, erosion of bridge foundations, sedimentation in lakes and harbors, coastal erosions, and continuing erosion and wasting of soil and landscapes can all negatively affect society.

UCP 3: Although most things are in the process of becoming different—changing—some properties of objects and processes are characterized by constancy, including the speed of light, the charge of an electron, and the total mass plus energy in the universe. Changes might occur, for example, in properties of materials, position of objects, motion, and form and function of systems. Interactions within and among systems result in change. Changes vary in rate, scale, and pattern, including trends and cycles.

FOCUS *(5 minutes)*

__ **Overview, TE** Review the objectives listed in the Student Edition. **(GENERAL)**

__ **PowerPoint® Resources** Use the customizable presentation to enhance your lecture.

__ **Bellringer, TE** Students write down information they would need if they were measuring an earthquake. **(GENERAL)**

__ **Bellringer Transparency** Use this transparency as students enter the classroom and find their seats.

MOTIVATE *(10 minutes)*

__ **Group Activity, Model Seismograph, TE** In groups, students use rolling pins and shoeboxes to build model seismographs. **(GENERAL)**

KEY **SE** = Student Edition **TE** = Teacher Edition **CRF** = Chapter Resource File

Lesson Plan

TEACH *(65 minutes)*

__ **Directed Reading, CRF** These worksheets reinforce basic concepts and vocabulary presented in the section. **(BASIC)**

__ **Quick Lab, Seismographic Record, SE** Students build model seismographs to learn how different materials affect vibrations. **(GENERAL)**

__ **Datasheet for Quick Lab, Seismographic Record, CRF** Students use the datasheet to complete the Quick Lab. **(GENERAL)**

__ **Reading Skill Builder, Reading Organizer, TE** Students create a reading organizer to help them review this section. **(BASIC)**

__ **Skill Builder, Graphing, TE** Students graph the seismic data of different earthquakes. **(ADVANCED)**

__ **Skills Practice Lab, Finding an Epicenter, SE** Students analyze an earthquake's P and S waves to determine the location of its epicenter. **(GENERAL)**

__ **Datasheet for Chapter Lab, Skills Practice Lab, Finding an Epicenter, CRF** Students use the datasheet to complete the Chapter Lab. **(GENERAL)**

__ **Career Focus, Geophysicist, SE** Geophysicists measure ground deformation to study how Earth's crust moves and to ultimately predict future events.

__ **Using the Figure, Discussion, TE** Students discuss how new satellite technology offers strategies for detecting stresses that could lead to earthquakes. **(GENERAL)**

CLOSE *(10 minutes)*

__ **Section Review, SE** Students answer end-of-section vocabulary, key ideas, critical thinking, and concept mapping questions. **(GENERAL)**

__ **Reteaching, Earthquake Posters, TE** Students draw pictures illustrating the increasing levels of damage earthquakes of different intensities create. **(BASIC)**

__ **Quiz, TE** Students answer 2 questions about this section. **(GENERAL)**

__ **Section Quiz, CRF** Use this quiz to assess students' understanding of the section. **(GENERAL) Also in Spanish**

__ **Spanish Section Quiz, Spanish Assessments** Use this quiz to assess students' understanding of the section. **(GENERAL)**

__ **Alternative Assessment, Sizing Up Quakes, TE** Students simulate earthquakes and design a scale to compare the magnitudes. **(ADVANCED)**

KEY **SE** = Student Edition **TE** = Teacher Edition **CRF** = Chapter Resource File

Section 3 Earthquakes and Society

PACING

Regular Schedule:	**with lab(s): N/A**	**without lab(s): 2 days**
Block Schedule:	**with lab(s): N/A**	**without lab(s): 1 day**

OBJECTIVES

1. **Discuss** the relationship between tsunamis and earthquakes.
2. **Describe** two possible effects of a major earthquake on buildings.
3. **List** three safety techniques to prevent injury caused by earthquake activity.
4. **Identify** four methods scientists use to forecast earthquake risks.

NATIONAL SCIENCE EDUCATION STANDARDS

PS 6: Interactions of energy and matter

SPSP 5: Natural and human induced hazards

SPSP 5a: Normal adjustments of earth may be hazardous for humans. Humans live at the interface between the atmosphere driven by solar energy and the upper mantle where convection creates changes in the earth's solid crust. As societies have grown, become stable, and come to value aspects of the environment, vulnerability to natural processes of change has increased.

SPSP 5c: Some hazards, such as earthquakes, volcanic eruptions, and severe weather, are rapid and spectacular. But there are slow and progressive changes that also result in problems for individuals and societies. For example, change in stream channel position, erosion of bridge foundations, sedimentation in lakes and harbors, coastal erosions, and continuing erosion and wasting of soil and landscapes can all negatively affect society.

ST 1: Abilities of Technological design

UCP 3: Although most things are in the process of becoming different—changing—some properties of objects and processes are characterized by constancy, including the speed of light, the charge of an electron, and the total mass plus energy in the universe. Changes might occur, for example, in properties of materials, position of objects, motion, and form and function of systems. Interactions within and among systems result in change. Changes vary in rate, scale, and pattern, including trends and cycles.

FOCUS *(5 minutes)*

__ **Overview, TE** Review the objectives listed in the Student Edition. **(GENERAL)**

__ **PowerPoint® Resources** Use the customizable presentation to enhance your lecture.

__ **Bellringer, TE** Students describe some effects of earthquakes. **(GENERAL)**

__ **Bellringer Transparency** Use this transparency as students enter the classroom and find their seats.

KEY **SE** = Student Edition **TE** = Teacher Edition **CRF** = Chapter Resource File

MOTIVATE *(10 minutes)*

__ **Demonstration, Giant Wave, TE** Students observe a model tsunami. **(GENERAL)**

TEACH *(20 minutes)*

__ **Directed Reading, CRF** These worksheets reinforce basic concepts and vocabulary presented in the section. **(BASIC)**

__ **Quick Lab, Earthquake-Safe Buildings, SE** Students build different kinds of structures and observe what happens to them during an "earthquake." **(GENERAL)**

__ **Datasheet for Quick Lab, Earthquake-Safe Buildings, CRF** Students use the datasheet to complete the Quick Lab. **(GENERAL)**

__ **Using the Figure, TE** Students discuss earthquake preparedness. **(BASIC)**

__ **Group Activity, Earthquake Safety, TE** In groups, students explore ways to stay safe during an earthquake. **(GENERAL)**

__ **Reading Skill Builder, Paired Summarizing, TE** Students read the text in pairs and take turns summarizing what they have read. **(BASIC)**

__ **Teaching Transparency 65, Seismic Gaps with Worksheet** Use this Teaching Transparency and the accompanying worksheet to support your classroom discussion of seismic gaps.

__ **Debate, Worth the Price? TE** Students research and debate the pros and cons of being able to predict when and where an earthquake will strike. **(ADVANCED)**

__ **Maps in Action, Earthquake Hazard Map, SE** Students enhance their map-reading skills by answering questions about a map showing the earthquake hazard levels of the world's continents. **(ADVANCED)**

__ **Teaching Transparency 66, Earthquake Hazard Map with Worksheet** Use this Teaching Transparency and the accompanying worksheet to help students perform the Maps in Action activity.

__ **Internet Activity, Hazards in the Americas, TE** Students plot recent earthquakes on a map. **(GENERAL)**

__ **Internet Activity, Hazards in the Americas, CRF** This worksheet directs student research for the topic presented in the Teacher Edition. **(GENERAL)**

__ **Skills Practice Lab, Earthquakes and Soil, CRF** Students model soil and simulate seismic waves to study the effects of liquefaction. **(GENERAL)**

__ **Interactive Tutor CD-ROM, Volcanoes and Earthquakes** Students practice what they have learned using this interactive activity. **(GENERAL)**

CLOSE *(10 minutes)*

__ **Section Review, SE** Students answer end-of-section vocabulary, key ideas, critical thinking, and concept mapping questions. **(GENERAL)**

__ **Reteaching, Questions, TE** Students compose questions about this section and play a quiz game to answer them. **(BASIC)**

KEY SE = Student Edition TE = Teacher Edition CRF = Chapter Resource File

__ **Quiz, TE** Students answer 3 questions about this section. **(GENERAL)**

__ **Section Quiz, CRF** Use this quiz to assess students' understanding of the section. **(GENERAL) Also in Spanish**

__ **Spanish Section Quiz, Spanish Assessments** Use this quiz to assess students' understanding of the section. **(GENERAL)**

__ **Alternative Assessment, Making Models, TE** Students construct dioramas illustrating the effects of earthquakes and how to stay safe during an earthquake. **(GENERAL)**

KEY **SE** = Student Edition **TE** = Teacher Edition **CRF** = Chapter Resource File

End of Chapter Review and Assessment

PACING

Regular Schedule: **with lab(s): N/A** **without lab(s): 2 days**
Block Schedule: **with lab(s): N/A** **without lab(s): 1 day**

__ **Chapter Review, SE** Students answer end-of-chapter vocabulary, key ideas, critical thinking, math, writing, and interpreting graphics questions. **(GENERAL)**

__ **Alternative Assessment, News Broadcast, TE** In groups, students perform news broadcasts describing the effects of an earthquake on a local community. **(GENERAL)**

__ **Standardized Test Prep, SE** Have students complete the standardized test prep to help them prepare for standardized tests. **(GENERAL)**

__ **Holt PuzzlePro®** Use this software to create crossword puzzles and word searches that make learning vocabulary fun. **(GENERAL)**

__ **Concept Review, CRF** Use this worksheet to assess student's understanding of the main concepts presented in the chapter. **(GENERAL)**

__ **Critical Thinking, CRF** Students use this worksheet to develop and practice critical thinking skills. **(ADVANCED)**

__ **Math Skills, CRF** This worksheet helps reinforce students' understanding of exponents. **(GENERAL)**

__ **Graphing Skills, CRF** This worksheet helps reinforce students' understanding of pie graphs. **(GENERAL)**

__ **Concept review, Study Guide, Minor Bodies of the Solar System** Use this worksheet to assess student's understanding of the main concepts presented in the chapter. **(GENERAL)**

__ **Concept Review, Spanish Study Guide, Minor Bodies of the Solar System** This worksheet helps reinforce English-language learners' understanding of the main concepts presented in the chapter. **(GENERAL)**

__ **Chapter Test A, CRF** Assign questions from the appropriate test for chapter assessment. **(GENERAL)**

__ **Chapter Test B, CRF** Assign questions from the appropriate test for chapter assessment. **(ADVANCED)**

__ **Spanish Chapter Test A, Spanish Assessments** Assign questions from the appropriate test for chapter assessment. **(GENERAL)**

__ **Test Generator, One-Stop Planner** Create a customized homework assignment, quiz, or test using the HRW Test Generator program. **(GENERAL)**

__ **Test Item Listing, CRF** Use the Test Item Listing to identify questions for a customized homework, quiz, or test.

__ **PowerPoint® Resources** Use the customizable presentation to help students practice standardized test-taking skills.

KEY **SE** = Student Edition **TE** = Teacher Edition **CRF** = Chapter Resource File

OTHER RESOURCE OPTIONS

__ **HRW Earth Science Video,** Segment 5, "Earthquakes" This segment explores the seismic waves generated by earthquakes, which travel at different speeds through the Earth's crust. **(GENERAL)**

__ **NOVA Video, "Earthquake"** This program shows how today's advanced technology helps geologists predict earthquakes.

__ **Brain Food Video Quiz** Use this video resource to assess student's progress. **(GENERAL)**

KEY **SE** = Student Edition **TE** = Teacher Edition **CRF** = Chapter Resource File

Chapter 13 Volcanoes

Chapter Opener

__ **Pre-Reading Activity, Table Fold, SE**

__ **Using the Figure, Forming Islands, TE (GENERAL)**

Section 1 Volcanoes and Plate Tectonics

PACING

Regular Schedule:	**with lab(s): N/A**	**without lab(s): 2 days**
Block Schedule:	**with lab(s): N/A**	**without lab(s): 1 day**

OBJECTIVES

1. **Describe** the three ways magma can form.
2. **Define** volcanism.
3. **Identify** three major volcanic zones.
4. **Describe** how magma can form plutons.

NATIONAL SCIENCE EDUCATION STANDARDS

ES 3c: Interactions among the solid earth, the oceans, the atmosphere, and organisms have resulted in the ongoing evolution of the earth system. We can observe some changes such as earthquakes and volcanic eruptions on a human time scale, but many processes such as mountain building and plate movements take place over hundreds of millions of years.

ST 1e: Communicate the problem, process, and solution. Students should present their results to students, teachers, and others in a variety of ways, such as orally, in writing, and in other forms—including models, diagrams, and demonstrations.

UCP 4: Evolution is a series of changes, some gradual and some sporadic, that accounts for the present form and function of objects, organisms, and natural and designed systems. The general idea of evolution is that the present arises from materials and forms of the past. Although evolution is most commonly associated with the biological theory explaining the process of descent with modification of organisms from common ancestors, evolution also describes changes in the universe.

FOCUS *(5 minutes)*

__ **Overview, TE** Review the objectives listed in the Student Edition. **(GENERAL)**

KEY **SE** = Student Edition **TE** = Teacher Edition **CRF** = Chapter Resource File

__ **PowerPoint® Resources** Use the customizable presentation to enhance your lecture.

__ **Bellringer, TE** Have students describe what they think of when they hear the terms volcano and volcanic eruption. **(GENERAL)**

__ **Bellringer Transparency** Use this transparency as students enter the classroom and find their seats.

MOTIVATE *(10 minutes)*

__ **Using the Figure, Pressure and Magma Formation, TE** Students use the figure showing temperature and pressure beneath Earth's surface to answer a question about Earth's mantle. **(GENERAL)**

TEACH *(20 minutes)*

__ **Directed Reading, CRF** These worksheets reinforce basic concepts and vocabulary presented in the section. **(BASIC)**

__ **Using the Figure, Volcanic Locations, TE** Students use the map of tectonic plate boundaries to determine where most of the active volcanoes are located. **(GENERAL)**

__ **Reading Skill Builder, Reading Organizer, TE** Students create concept maps about magma. **(BASIC)**

__ **Skill Builder, Vocabulary, TE** Students review the meanings of the words *volcano* and *lava*. **(GENERAL)**

__ **Teaching Transparency 67, Volcanoes and Tectonic Plate Boundaries with Worksheet** Use this Teaching Transparency and the accompanying worksheet to support your classroom discussion of volcanoes and tectonic plate boundaries.

__ **Quick Lab, Changing Melting Point, SE** Students use ice, paper plates, and salt to observe how ice melts. **(GENERAL)**

__ **Datasheet for Quick Lab, Changing Melting Point, CRF** Students use the datasheet to complete the Quick Lab. **(GENERAL)**

__ **Social Studies Connection, Geothermal Energy, TE** Students research instances where volcanic activity has been utilized for public benefit. **(BASIC)**

__ **Skill Builder, Graphing, TE** Students research and graph the locations and dates of volcanic activity in Yellowstone National Park. **(ADVANCED)**

__ **Teaching Transparency 68, Hot Spots and Mantle Plumes with Worksheet** Use this Teaching Transparency and the accompanying worksheet to support your classroom discussion of hot spots and mantle plumes.

__ **Maps in Action, The Hawaiian-Emperor Seamount Chain, SE** Students enhance their map-reading skills by answering questions about a map showing the locations and ages of islands and seamounts in the Hawaiian-Emperor seamount chain. **(GENERAL)**

__ **Teaching Transparency 70, The Hawaiian-Emperor Seamount Chain with Worksheet** Use this Teaching Transparency and the accompanying worksheet to help students perform the Maps in Action activity.

KEY **SE** = Student Edition **TE** = Teacher Edition **CRF** = Chapter Resource File

__ **Group Activity, Charting Volcanic Activity, TE** In groups, students study a geophysical map showing long-term volcanic activity. **(ADVANCED)**

__ **Making Models Lab, Magma in Earth's Crust, CRF** Students model how magma rises through Earth's crust and identify how pressure affects its ability to rise. **(GENERAL)**

CLOSE *(10 minutes)*

__ **Section Review, SE** Students answer end-of-section vocabulary, key ideas, critical thinking, and concept mapping questions. **(GENERAL)**

__ **Reteaching, Volcanic Settings, TE** Students review tectonic plates. **(BASIC)**

__ **Quiz, TE** Students answer 3 questions about this section. **(GENERAL)**

__ **Section Quiz, CRF** Use this quiz to assess students' understanding of the section. **(GENERAL) Also in Spanish**

__ **Spanish Section Quiz, Spanish Assessments** Use this quiz to assess students' understanding of the section. **(GENERAL)**

__ **Alternative Assessment, Modeling Volcanic Processes, TE** Students create models showing a volcano's various stages of formation. **(ADVANCED)**

KEY **SE** = Student Edition **TE** = Teacher Edition **CRF** = Chapter Resource File

Lesson Plan

Section 2 Volcanic Eruptions

PACING

Regular Schedule:	**with lab(s): 2 days**	**without lab(s): 1 day**
Block Schedule:	**with lab(s): 1 day**	**without lab(s): 0.5 day**

OBJECTIVES

1. **Explain** how the composition of magma affects volcanic eruptions and lava flow.
2. **Describe** the five major types of pyroclastic material.
3. **Identify** the three main types of volcanic cones.
4. **Describe** how a caldera forms.
5. **List** three events that may signal a volcanic eruption.

NATIONAL SCIENCE EDUCATION STANDARDS

ES 3c: Interactions among the solid earth, the oceans, the atmosphere, and organisms have resulted in the ongoing evolution of the earth system. We can observe some changes such as earthquakes and volcanic eruptions on a human time scale, but many processes such as mountain building and plate movements take place over hundreds of millions of years.

SPSP 5a: Normal adjustments of earth may be hazardous for humans. Humans live at the interface between the atmosphere driven by solar energy and the upper mantle where convection creates changes in the earth's solid crust. As societies have grown, become stable, and come to value aspects of the environment, vulnerability to natural processes of change has increased.

ST 1e: Communicate the problem, process, and solution. Students should present their results to students, teachers, and others in a variety of ways, such as orally, in writing, and in other forms—including models, diagrams, and demonstrations.

UCP 2: Evidence consists of observations and data on which to base scientific explanations. Using evidence to understand interactions allows individuals to predict changes in natural and designed systems. Models are tentative schemes or structures that correspond to real objects, events, or classes of events, and that have explanatory power. Models help scientists and engineers understand how things work. Models take many forms, including physical objects, plans, mental constructs, mathematical equations, and computer simulations.

KEY SE = Student Edition TE = Teacher Edition **CRF** = Chapter Resource File

Lesson Plan

FOCUS *(5 minutes)*

__ **Overview, TE** Review the objectives listed in the Student Edition. **(GENERAL)**

__ **PowerPoint® Resources** Use the customizable presentation to enhance your lecture.

__ **Bellringer, TE** Start a discussion about the characteristics of a volcanic eruption by having students draw a picture of an eruption. **(GENERAL)**

__ **Bellringer Transparency** Use this transparency as students enter the classroom and find their seats.

MOTIVATE *(10 minutes)*

__ **Demonstration, Viscosity, TE** Using a straw, blow bubbles into beakers of water, cooking oil, and honey to demonstrate how viscous liquids trap more gases. **(GENERAL)**

TEACH *(65 minutes)*

__ **Directed Reading, CRF** These worksheets reinforce basic concepts and vocabulary presented in the section. **(BASIC)**

__ **Using the Figure, Lava Flows, TE** Students use the figure to review the names of mafic magma. **(BASIC)**

__ **Graphic Organizer, Spider Map, SE** Use the graphic organizer to help students develop their analytical skills. **(GENERAL)**

__ **Activity, Poster Project, TE** Students research the classification system for volcanic explosivity (VEI). **(ADVANCED)**

__ **Art Connection, The Scream, TE** Students research works of art that may have been influenced by volcanoes or volcanic eruptions. **(GENERAL)**

__ **Using the Figure, Volcano Types, TE** Students use the figure showing the three types of volcanic cones to review characteristics of types of volcanoes. **(GENERAL)**

__ **Reading Skill Builder, Reading Hint, TE** Help students to recall which types of lavas form each type of volcano. **(BASIC)**

__ **Teaching Transparency 69, Types of Volcanoes with Worksheet** Use this Teaching Transparency and the accompanying worksheet to support your classroom discussion of types of volcanoes.

__ **Astronomy Connection, Extraterrestrial Volcanoes, TE** Students research volcanic activity on other planets and moons and present their findings. **(ADVANCED)**

__ **Group Activity, Historic Eruptions, TE** In groups, students research a volcanic eruption. **(GENERAL)**

__ **Quick Lab, Volcanic Cones, SE** Students build models of a cinder cone and a shield volcano. **(GENERAL)**

__ **Datasheet for Quick Lab, Volcanic Cones, CRF** Students use the datasheet to complete the Quick Lab. **(GENERAL)**

__ **Making Models Lab, Volcano Verdict, SE** Students measure the carbon dioxide emitted from a simulated volcano. **(GENERAL)**

KEY SE = Student Edition TE = Teacher Edition CRF = Chapter Resource File

__ **Datasheet for Chapter Lab, Making Models Lab, Volcano Verdict, CRF** Students use the datasheet to complete the Chapter Lab. **(GENERAL)**

__ **Eye on the Environment, The Effects of Volcanoes on Climate, SE** Scientists study the effects of volcanoes on global climate to learn how the atmosphere is affected by both natural events and human sources.

__ **Internet Activity, The Carbon Cycle, TE** Students research the carbon cycle and present their findings. **(ADVANCED)**

__ **Internet Activity, The Carbon Cycle, CRF** This worksheet directs student research for the topic presented in the Teacher Edition. **(GENERAL)**

__ **Inquiry Lab, Lava Flows, CRF** Students examine how viscosity affects the rate at which a fluid flows and identify factors affecting viscosity. **(GENERAL)**

__ **Interactive Tutor CD-ROM, Volcanoes and Earthquakes** Students practice what they have learned using this interactive activity. **(GENERAL)**

CLOSE *(10 minutes)*

__ **Section Review, SE** Students answer end-of-section vocabulary, key ideas, critical thinking, and concept mapping questions. **(GENERAL)**

__ **Reteaching, Volcanic, Magma, and Eruptions, TE** Students make vocabulary cards to review descriptions of volcanoes. **(BASIC)**

__ **Quiz, TE** Students answer 2 questions about this section. **(GENERAL)**

__ **Section Quiz, CRF** Use this quiz to assess students' understanding of the section. **(GENERAL) Also in Spanish**

__ **Spanish Section Quiz, Spanish Assessments** Use this quiz to assess students' understanding of the section. **(GENERAL)**

__ **Alternative Assessment, Modeling Volcano Formation, TE** Students build models showing either volcano formation or caldera collapse. **(ADVANCED)**

KEY **SE** = Student Edition **TE** = Teacher Edition **CRF** = Chapter Resource File

Lesson Plan

End of Chapter Review and Assessment

PACING

Regular Schedule: **with lab(s): N/A** **without lab(s): 2 days**
Block Schedule: **with lab(s): N/A** **without lab(s): 1 day**

__ **Chapter Review, SE** Students answer end-of-chapter vocabulary, key ideas, critical thinking, math, writing, and interpreting graphics questions. **(GENERAL)**

__ **Alternative Assessment, Chain-of-Events Charts, TE** Students create three chain-of-events charts to represent each type of volcano. **(GENERAL)**

__ **Standardized Test Prep, SE** Have students complete the standardized test prep to help them prepare for standardized tests. **(GENERAL)**

__ **Holt PuzzlePro®** Use this software to create crossword puzzles and word searches that make learning vocabulary fun. **(GENERAL)**

__ **Concept Review, CRF** Use this worksheet to assess student's understanding of the main concepts presented in the chapter. **(GENERAL)**

__ **Critical Thinking, CRF** Students use this worksheet to develop and practice critical thinking skills. **(ADVANCED)**

__ **Math Skills, CRF** This worksheet helps reinforce students' understanding of exponents. **(GENERAL)**

__ **Graphing Skills, CRF** This worksheet helps reinforce students' understanding of line graphs. **(GENERAL)**

__ **Concept Review, Study Guide, Volcanoes** Use this worksheet to assess student's understanding of the main concepts presented in the chapter. **(GENERAL)**

__ **Concept Review, Spanish Study Guide, Volcanoes** This worksheet helps reinforce English-language learners' understanding of the main concepts presented in the chapter. **(GENERAL)**

__ **Chapter Test A, CRF** Assign questions from the appropriate test for chapter assessment. **(GENERAL)**

__ **Chapter Test B, CRF** Assign questions from the appropriate test for chapter assessment. **(ADVANCED)**

__ **Spanish Chapter Test A, Spanish Assessments** Assign questions from the appropriate test for chapter assessment. **(GENERAL)**

__ **Test Generator, One-Stop Planner** Create a customized homework assignment, quiz, or test using the HRW Test Generator program. **(GENERAL)**

__ **Test Item Listing, CRF** Use the Test Item Listing to identify questions for a customized homework, quiz, or test.

__ **PowerPoint® Resources** Use the customizable presentation to help students practice standardized test-taking skills.

KEY **SE** = Student Edition **TE** = Teacher Edition **CRF** = Chapter Resource File

OTHER RESOURCE OPTIONS

__ **HRW Earth Science Video, Segment 6, "Volcanoes"** This segment examines volcanism at convergent and divergent plate boundaries. **(GENERAL)**

__ **Brain Food Video Quiz** Use this video resource to assess student's progress. **(GENERAL)**

KEY **SE** = Student Edition **TE** = Teacher Edition **CRF** = Chapter Resource File

Chapter 14 Weathering and Erosion

Chapter Opener

__ **Pre-Reading Activity, TriFold, SE**

__ **Using the Figure, Uluru, TE (GENERAL)**

Section 1 Weathering Processes

PACING

Regular Schedule:	**with lab(s): N/A**	**without lab(s): 2 days**
Block Schedule:	**with lab(s): N/A**	**without lab(s): 1 day**

OBJECTIVES

1. **Identify** three agents of mechanical weathering.
2. **Compare** mechanical and chemical weathering processes.
3. **Describe** four chemical reactions that decompose rock.

NATIONAL SCIENCE EDUCATION STANDARDS

ES 3c: Interactions among the solid earth, the oceans, the atmosphere, and organisms have resulted in the ongoing evolution of the earth system. We can observe some changes such as earthquakes and volcanic eruptions on a human time scale, but many processes such as mountain building and plate movements take place over hundreds of millions of years.

UPC 3: Although most things are in the process of becoming different—changing—some properties of objects and processes are characterized by constancy, including the speed of light, the charge of an electron, and the total mass plus energy in the universe. Changes might occur, for example, in properties of materials, position of objects, motion, and form and function of systems. Interactions within and among systems result in change. Changes vary in rate, scale, and pattern, including trends and cycles.

UPC 4: Evolution is a series of changes, some gradual and some sporadic, that accounts for the present form and function of objects, organisms, and natural and designed systems. The general idea of evolution is that the present arises from materials and forms of the past. Although evolution is most commonly associated with the biological theory explaining the process of descent with modification of organisms from common ancestors, evolution also describes changes in the universe.

FOCUS *(5 minutes)*

__ **Overview, TE** Review the objectives listed in the Student Edition. **(GENERAL)**

__ **PowerPoint® Resources** Use the customizable presentation to enhance your lecture.

__ **Bellringer, TE** Students list natural agents that could break rock. **(GENERAL)**

KEY **SE** = Student Edition **TE** = Teacher Edition **CRF** = Chapter Resource File

__ **Bellringer Transparency** Use this transparency as students enter the classroom and find their seats.

MOTIVATE *(10 minutes)*

__ **Activity, Weathering Walk, TE** Students list examples of weathering found in their neighborhoods. **(GENERAL)**

TEACH *(20 minutes)*

__ **Directed Reading, CRF** These worksheets reinforce basic concepts and vocabulary presented in the section. **(BASIC)**

__ **Activity, Ice Wedging, TE** Students freeze clay balls and observe the weathering changes that take place. **(GENERAL)**

__ **Reading Skill Builder, Paired Summarizing, TE** Students read the text in pairs and take turns summarizing what they have read. **(BASIC)**

__ **Quick Lab, Mechanical Weathering, SE** Students use water and rock chips to explore mechanical weathering. **(GENERAL)**

__ **Datasheet for Quick Lab, Mechanical Weathering, CRF** Students use the datasheet to complete the Quick Lab. **(GENERAL)**

__ **Group Activity, Finding Evidence, TE** Students look for examples of weathering at a local park. **(GENERAL)**

__ **Demonstration, Rusting, TE** Students brainstorm ways to prevent oxidation. **(GENERAL)**

__ **Discussion, Weathering on the Moon, TE** Students discuss whether the moon has been mechanically or chemically weathered. **(GENERAL)**

__ **Activity, Chalk Sculptures, TE** Students study the effects of vinegar on chalk in to help them understand acid rain. **(GENERAL)**

__ **Teaching Transparency 71, Chemical Weathering with Worksheet** Use this Teaching Transparency and the accompanying worksheet to support your classroom discussion of chemical weathering.

__ **Interactive Tutor CD-ROM, Physical and Chemical Weathering** Students practice what they have learned using this interactive activity. **(GENERAL)**

CLOSE *(10 minutes)*

__ **Section Review, SE** Students answer end-of-section vocabulary, key ideas, critical thinking, and concept mapping questions. **(GENERAL)**

__ **Reteaching, Weathering Processes, TE** Students use flashcards to study mechanical and chemical weathering processes. **(BASIC)**

__ **Quiz, TE** Students answer 2 questions about this section. **(GENERAL)**

__ **Section Quiz, CRF** Use this quiz to assess students' understanding of the section. **(GENERAL) Also in Spanish**

KEY **SE** = Student Edition **TE** = Teacher Edition **CRF** = Chapter Resource File

Lesson Plan

__ **Spanish Section Quiz, Spanish Assessments** Use this quiz to assess students' understanding of the section. **(GENERAL)**

__ **Alternative Assessment, Table, TE** Students create tables comparing mechanical and chemical weathering. **(GENERAL)**

KEY SE = Student Edition **TE** = Teacher Edition **CRF** = Chapter Resource File

Lesson Plan

Section 2 Rates of Weathering

PACING

Regular Schedule:	**with lab(s): N/A**	**without lab(s): 2 days**
Block Schedule:	**with lab(s): N/A**	**without lab(s): 1 day**

OBJECTIVES

1. **Explain** how rock composition affects the rate of weathering.
2. **Discuss** how surface area affects the rate at which rock weathers.
3. **Describe** the effects of climate and topography on the rate of weathering.

NATIONAL SCIENCE EDUCATION STANDARDS

ES 3c: Interactions among the solid earth, the oceans, the atmosphere, and organisms have resulted in the ongoing evolution of the earth system. We can observe some changes such as earthquakes and volcanic eruptions on a human time scale, but many processes such as mountain building and plate movements take place over hundreds of millions of years.

SPSP 4a: Natural ecosystems provide an array of basic processes that affect humans. Those processes include maintenance of the quality of the atmosphere, generation of soils, control of the hydrologic cycle, disposal of wastes, and recycling of nutrients. Humans are changing many of these basic processes, and the changes may be detrimental to humans.

UPC 3: Although most things are in the process of becoming different—changing—some properties of objects and processes are characterized by constancy, including the speed of light, the charge of an electron, and the total mass plus energy in the universe. Changes might occur, for example, in properties of materials, position of objects, motion, and form and function of systems. Interactions within and among systems result in change. Changes vary in rate, scale, and pattern, including trends and cycles.

UPC 4: Evolution is a series of changes, some gradual and some sporadic, that accounts for the present form and function of objects, organisms, and natural and designed systems. The general idea of evolution is that the present arises from materials and forms of the past. Although evolution is most commonly associated with the biological theory explaining the process of descent with modification of organisms from common ancestors, evolution also describes changes in the universe.

FOCUS *(5 minutes)*

__ **Overview, TE** Review the objectives listed in the Student Edition. **(GENERAL)**

__ **PowerPoint® Resources** Use the customizable presentation to enhance your lecture.

__ **Bellringer, TE** Students discuss factors that affect weathering. **(GENERAL)**

KEY **SE** = Student Edition **TE** = Teacher Edition **CRF** = Chapter Resource File

Lesson Plan

__ **Bellringer Transparency** Use this transparency as students enter the classroom and find their seats.

MOTIVATE *(10 minutes)*

__ **Internet Activity, National Parks, TE** Students research the role of differential weathering on the formation of a national park's landforms. **(GENERAL)**

__ **Internet Activity, National Parks, CRF** This worksheet directs student research for the topic presented in the Teacher Edition. **(GENERAL)**

TEACH *(20 minutes)*

__ **Directed Reading, CRF** These worksheets reinforce basic concepts and vocabulary presented in the section. **(BASIC)**

__ **Quick Lab, Surface Areas, SE** Students use sugar cubes and water to examine the relationship between surface area and weathering. **(GENERAL)**

__ **Datasheet for Quick Lab, Surface Areas, CRF** Students use the datasheet to complete the Quick Lab. **(GENERAL)**

__ **Teaching Transparency 72, Surface Area with Worksheet** Use this Teaching Transparency and the accompanying worksheet to support your classroom discussion of surface area.

__ **Activity, Create a Cave, TE** Students use sand and sugar cubes to create miniature caverns. **(GENERAL)**

CLOSE *(10 minutes)*

__ **Section Review, SE** Students answer end-of-section vocabulary, key ideas, critical thinking, and concept mapping questions. **(GENERAL)**

__ **Homework, Research Paper, TE** Students research efforts to preserve and restore famous monuments. **(GENERAL)**

__ **Reteaching, Peer Reviewing, TE** In pairs, students compose questions about this section and quiz each other. **(BASIC)**

__ **Quiz, TE** Students answer 2 questions about this section. **(GENERAL)**

__ **Section Quiz, CRF** Use this quiz to assess students' understanding of the section. **(GENERAL) Also in Spanish**

__ **Spanish Section Quiz, Spanish Assessments** Use this quiz to assess students' understanding of the section. **(GENERAL)**

__ **Alternative Assessment, Concept Collage, TE** Students use pictures of differential weathering to create a concept map of this section. **(GENERAL)**

KEY SE = Student Edition TE = Teacher Edition CRF = Chapter Resource File

Lesson Plan

Section 3 Soil

PACING

Regular Schedule:	**with lab(s): 2 days**	**without lab(s): 1 day**
Block Schedule:	**with lab(s): 1 day**	**without lab(s): 0.5 day**

OBJECTIVES

1. **Summarize** how soils form.
2. **Explain** how the composition of parent rock affects soil composition.
3. **Describe** the characteristic layers of mature residual soils.
4. **Predict** the type of soil that will form in arctic and tropical climates.

NATIONAL SCIENCE EDUCATION STANDARDS

ES 3: Origin and evolution of the Earth system

SPSP 4a: Natural ecosystems provide an array of basic processes that affect humans. Those processes include maintenance of the quality of the atmosphere, generation of soils, control of the hydrologic cycle, disposal of wastes, and recycling of nutrients. Humans are changing many of these basic processes, and the changes may be detrimental to humans.

UPC 3: Although most things are in the process of becoming different—changing—some properties of objects and processes are characterized by constancy, including the speed of light, the charge of an electron, and the total mass plus energy in the universe. Changes might occur, for example, in properties of materials, position of objects, motion, and form and function of systems. Interactions within and among systems result in change. Changes vary in rate, scale, and pattern, including trends and cycles.

UPC 4: Evolution is a series of changes, some gradual and some sporadic, that accounts for the present form and function of objects, organisms, and natural and designed systems. The general idea of evolution is that the present arises from materials and forms of the past. Although evolution is most commonly associated with the biological theory explaining the process of descent with modification of organisms from common ancestors, evolution also describes changes in the universe.

FOCUS *(5 minutes)*

___ **Overview, TE** Review the objectives listed in the Student Edition. **(GENERAL)**

___ **PowerPoint® Resources** Use the customizable presentation to enhance your lecture.

___ **Bellringer, TE** Students define the word *soil.* **(GENERAL)**

___ **Bellringer Transparency** Use this transparency as students enter the classroom and find their seats.

KEY **SE** = Student Edition **TE** = Teacher Edition **CRF** = Chapter Resource File

Lesson Plan

MOTIVATE *(10 minutes)*

__ **Group Activity, What is Soil? TE** Students sort and analyze soil samples. **(GENERAL)**

TEACH *(65 minutes)*

__ **Directed Reading, CRF** These worksheets reinforce basic concepts and vocabulary presented in the section. **(BASIC)**

__ **Using the Figure, Soil Horizons, TE** Students study the figure of soil horizons and create a profile of the school's soil. **(GENERAL)**

__ **Teaching Transparency 73, Soil Horizons of Residual Soils with Worksheet** Use this Teaching Transparency and the accompanying worksheet to support your classroom discussion of soil horizons of residual soils.

__ **Skill Builder, Math, TE** Students calculate the range of rates of soil formation. **(BASIC)**

__ **Skills Practice Lab, Soil Chemistry, SE** Students use the pH scale to test the acidity of soil samples. **(GENERAL)**

__ **Datasheet for Chapter Lab, Skills Practice Lab, Soil Chemistry, CRF** Students use the datasheet to complete the Chapter Lab. **(GENERAL)**

__ **Maps in Action, Soil Map of North Carolina, SE** Students enhance their map-reading skills by answering questions about a map showing soil systems in North Carolina. **(GENERAL)**

__ **Teaching Transparency with Worksheet, Soil Map of North Carolina** Use this Teaching Transparency and the accompanying worksheet to help students perform the Maps in Action activity.

__ **Discussion, Local Soils, TE** Students discuss the types of soil found in their state or region. **(GENERAL)**

__ **Inquiry Lab, Acid Rain and Soils, CRF** Students observe how different types of soil can alter the pH of water draining through them. **(GENERAL)**

__ **Making Models Lab, Soil Profiles, CRF** Students build a model soil profile to understand how soil forms. **(GENERAL)**

CLOSE *(10 minutes)*

__ **Section Review, SE** Students answer end-of-section vocabulary, key ideas, critical thinking, and concept mapping questions. **(GENERAL)**

__ **Reteaching, Flipbook, TE** Students create flipbooks showing the different types of soil. **(BASIC)**

__ **Quiz, TE** Students answer 2 questions about this section. **(GENERAL)**

__ **Section Quiz, CRF** Use this quiz to assess students' understanding of the section. **(GENERAL) Also in Spanish**

__ **Spanish Section Quiz, Spanish Assessments** Use this quiz to assess students' understanding of the section. **(GENERAL)**

KEY **SE** = Student Edition **TE** = Teacher Edition **CRF** = Chapter Resource File

Lesson Plan

__ **Alternative Assessment, Soil Profiles, TE** Students create model soil profiles of a climate and write paragraphs describing its formation. **(GENERAL)**

KEY **SE** = Student Edition **TE** = Teacher Edition **CRF** = Chapter Resource File

Section 4 Erosion

PACING

Regular Schedule: **with lab(s): N/A** **without lab(s): 2 days**
Block Schedule: **with lab(s): N/A** **without lab(s): 1 day**

OBJECTIVES

1. **Define** erosion, and list four agents of erosion.
2. **Identify** four farming methods that conserve soil.
3. **Discuss** two ways gravity contributes to erosion.
4. **Describe** the three major landforms shaped by weathering and erosion.

NATIONAL SCIENCE EDUCATION STANDARDS

ES 3c: Interactions among the solid earth, the oceans, the atmosphere, and organisms have resulted in the ongoing evolution of the earth system. We can observe some changes such as earthquakes and volcanic eruptions on a human time scale, but many processes such as mountain building and plate movements take place over hundreds of millions of years.

SPSP 4a: Natural ecosystems provide an array of basic processes that affect humans. Those processes include maintenance of the quality of the atmosphere, generation of soils, control of the hydrologic cycle, disposal of wastes, and recycling of nutrients. Humans are changing many of these basic processes, and the changes may be detrimental to humans.

SPSP 5c: Some hazards, such as earthquakes, volcanic eruptions, and severe weather, are rapid and spectacular. But there are slow and progressive changes that also result in problems for individuals and societies. For example, change in stream channel position, erosion of bridge foundations, sedimentation in lakes and harbors, coastal erosions, and continuing erosion and wasting of soil and landscapes can all negatively affect society.

UPC 3: Although most things are in the process of becoming different—changing—some properties of objects and processes are characterized by constancy, including the speed of light, the charge of an electron, and the total mass plus energy in the universe. Changes might occur, for example, in properties of materials, position of objects, motion, and form and function of systems. Interactions within and among systems result in change. Changes vary in rate, scale, and pattern, including trends and cycles.

UPC 4: Evolution is a series of changes, some gradual and some sporadic, that accounts for the present form and function of objects, organisms, and natural and designed systems. The general idea of evolution is that the present arises from materials and forms of the past. Although evolution is most commonly associated with the biological theory explaining the process of descent with modification of organisms from common ancestors, evolution also describes changes in the universe.

KEY **SE** = Student Edition **TE** = Teacher Edition **CRF** = Chapter Resource File

Lesson Plan

FOCUS *(5 minutes)*

__ **Overview, TE** Review the objectives listed in the Student Edition. **(GENERAL)**

__ **PowerPoint® Resources** Use the customizable presentation to enhance your lecture.

__ **Bellringer, TE** Students discuss why the Rocky Mountains are different from the Appalachians. **(GENERAL)**

__ **Bellringer Transparency** Use this transparency as students enter the classroom and find their seats.

MOTIVATE *(10 minutes)*

__ **Using the Figure, Local Vulnerability, TE** Students study a map of soil vulnerability and determine the vulnerability of their area. **(GENERAL)**

TEACH *(20 minutes)*

__ **Directed Reading, CRF** These worksheets reinforce basic concepts and vocabulary presented in the section. **(BASIC)**

__ **Teaching Transparency 74, Soil Erosion Vulnerability Map with Worksheet** Use this Teaching Transparency and the accompanying worksheet to support your classroom discussion of soil erosion vulnerability.

__ **History Connection, The Dust Bowl, TE** Students read excerpts from books to better understand the impact of soil erosion. **(GENERAL)**

__ **Skill Builder, Math, TE** Students calculate how much soil could be lost due to erosion in a 20-year period. **(GENERAL)**

__ **Graphic Organizer, Spider Map, SE** Use the graphic organizer to help students develop their analytical skills. **(GENERAL)**

__ **Demonstration, Soil Conservation, TE** Students and model gullying and terracing to review erosion. **(BASIC)**

__ **Internet Activity, NRCS, TE** Students research soil conservation programs in their area. **(GENERAL)**

__ **Internet Activity, NRCS, CRF** This worksheet directs student research for the topic presented in the Teacher Edition. **(GENERAL)**

__ **Activity, Plants and Soil, TE** Students observe how roots help stop erosion. **(BASIC)**

__ **Social Studies Connection, The Hills of Southern California, TE** Students debate the value of building in areas prone to mass movements. **(GENERAL)**

__ **Skill Builder, Writing, TE** Students research a recent mass movement and write a newspaper article describing the event. **(ADVANCED)**

__ **Discussion, Protecting Against Mass Movements, TE** Students discuss signs of mass movement and ways to combat it. **(GENERAL)**

__ **Reading Skill Builder, Reading Organizer, TE** Students are encouraged to take notes while reading this section. **(BASIC)**

__ **Demonstration, Solifluction, TE** Students observe and discuss solifluction. **(BASIC)**

KEY **SE** = Student Edition **TE** = Teacher Edition **CRF** = Chapter Resource File

__ **Career Focus, Soil Conservationist, SE** The work of a soil conservationist is detailed in this article.

__ **Activity, Guest Speaker, TE** Invite someone from a local soil conservation district to speak about local soil erosion problems. **(GENERAL)**

__ **Interactive Tutor CD-ROM, Soils** Students practice what they have learned using this interactive activity. **(GENERAL)**

CLOSE *(10 minutes)*

__ **Section Review, SE** Students answer end-of-section vocabulary, key ideas, critical thinking, and concept mapping questions. **(GENERAL)**

__ **Reteaching, Section Highlights, TE** Students are quizzed after summarizing the main points of the section. **(BASIC)**

__ **Quiz, TE** Students answer 2 questions about this section. **(GENERAL)**

__ **Section Quiz, CRF** Use this quiz to assess students' understanding of the section. **(GENERAL) Also in Spanish**

__ **Spanish Section Quiz, Spanish Assessments** Use this quiz to assess students' understanding of the section. **(GENERAL)**

__ **Alternative Assessment, Conservation Plans, TE** In groups, students develop a conservation plan to reduce weathering and erosion in their area. **(GENERAL)**

KEY **SE** = Student Edition **TE** = Teacher Edition **CRF** = Chapter Resource File

End of Chapter Review and Assessment

PACING

Regular Schedule:	**with lab(s): N/A**	**without lab(s): 2 days**
Block Schedule:	**with lab(s): N/A**	**without lab(s): 1 day**

__ **Chapter Review, SE** Students answer end-of-chapter vocabulary, key ideas, critical thinking, math, writing, and interpreting graphics questions. **(GENERAL)**

__ **Alternative Assessment, Life as a Rock, TE** Students write the life story of a rock. **(ADVANCED)**

__ **Standardized Test Prep, SE** Have students complete the standardized test prep to help them prepare for standardized tests. **(GENERAL)**

__ **Holt PuzzlePro®** Use this software to create crossword puzzles and word searches that make learning vocabulary fun. **(GENERAL)**

__ **Concept Review, CRF** Use this worksheet to assess student's understanding of the main concepts presented in the chapter. **(GENERAL)**

__ **Critical Thinking, CRF** Students use this worksheet to develop and practice critical thinking skills. **(ADVANCED)**

__ **Math Skills, CRF** This worksheet helps reinforce students' understanding of algebraic rearrangements. **(GENERAL)**

__ **Graphing Skills, CRF** This worksheet helps reinforce students' understanding of line graphs. **(GENERAL)**

__ **Study Guide, Concept Review, Weathering and Erosion** Use this worksheet to assess student's understanding of the main concepts presented in the chapter. **(GENERAL)**

__ **Concept Review, Spanish Study Guide, Weathering and Erosion** This worksheet helps reinforce English-language learners' understanding of the main concepts presented in the chapter. **(GENERAL)**

__ **Chapter Test A, CRF** Assign questions from the appropriate test for chapter assessment. **(GENERAL)**

__ **Chapter Test B, CRF** Assign questions from the appropriate test for chapter assessment. **(ADVANCED)**

__ **Spanish Chapter Test A, Spanish Assessments** Assign questions from the appropriate test for chapter assessment. **(GENERAL)**

__ **Test Generator, One-Stop Planner** Create a customized homework assignment, quiz, or test using the HRW Test Generator program. **(GENERAL)**

__ **Test Item Listing, CRF** Use the Test Item Listing to identify questions for a customized homework, quiz, or test.

__ **PowerPoint® Resources** Use the customizable presentation to help students practice standardized test-taking skills.

KEY **SE** = Student Edition **TE** = Teacher Edition **CRF** = Chapter Resource File

OTHER RESOURCE OPTIONS

__ **HRW Earth Science Video, Segment 9, "Weathering and Erosion"** Weathering and erosion, which continually reshape the landscape of Earth, are the focuses of this segment.

KEY **SE** = Student Edition **TE** = Teacher Edition **CRF** = Chapter Resource File

Chapter 15 River Systems

Chapter Opener

__ **Pre-Reading Activity, Layered Book, SE**

__ **Using the Figure, Developing Wilderness, TE (GENERAL)**

__ **Long-Term Project, Water Clarity, Long-Term Projects** Students use a Secchi disc to measure and monitor the clarity of a body of water for one month. **(ADVANCED)**

__ **Long-Term Project, Precipitation and the Water Table, Long-Term Projects** Students discover how the level of the water table varies with the amount of precipitation. **(ADVANCED)**

Section 1 The Water Cycle

PACING

Regular Schedule:	**with lab(s): N/A**	**without lab(s): 2 days**
Block Schedule:	**with lab(s): N/A**	**without lab(s): 1 day**

OBJECTIVES

1. **Outline** the stages of the water cycle.
2. **Describe** the factors that affect a water budget.
3. **List** two approaches to water conservation.

NATIONAL SCIENCE EDUCATION STANDARDS

ES 3c: Interactions among the solid earth, the oceans, the atmosphere, and organisms have resulted in the ongoing evolution of the earth system. We can observe some changes such as earthquakes and volcanic eruptions on a human time scale, but many processes such as mountain building and plate movements take place over hundreds of millions of years.

SPSP 3c: Humans use many natural systems as resources. Natural systems have the capacity to reuse waste, but that capacity is limited. Natural systems can change to an extent that exceeds the limits of organisms to adapt naturally or humans to adapt technologically.

UCP 4: Evolution is a series of changes, some gradual and some sporadic, that accounts for the present form and function of objects, organisms, and natural and designed systems. The general idea of evolution is that the present arises from materials and forms of the past. Although evolution is most commonly associated with the biological theory

KEY **SE** = Student Edition **TE** = Teacher Edition **CRF** = Chapter Resource File

explaining the process of descent with modification of organisms from common ancestors, evolution also describes changes in the universe.

FOCUS *(5 minutes)*

__ **Overview, TE** Review the objectives listed in the Student Edition. **(GENERAL)**

__ **PowerPoint® Resources** Use the customizable presentation to enhance your lecture.

__ **Bellringer, TE** Have students list as many properties and characteristics of water as they can think of. **(GENERAL)**

__ **Bellringer Transparency** Use this transparency as students enter the classroom and find their seats.

MOTIVATE *(10 minutes)*

__ **Group Activity, Modeling the Water Cycle, TE** In groups, students build a terrarium to model the water cycle. **(GENERAL)**

TEACH *(20 minutes)*

__ **Directed Reading, CRF** These worksheets reinforce basic concepts and vocabulary presented in the section. **(BASIC)**

__ **Skill Builder, Math, TE** Students calculate the transpiration rate of a plant. **(GENERAL)**

__ **Reading Skill Builder, Paired Summarizing, TE** Students read the text in pairs and take turns summarizing what they have read. **(BASIC)**

__ **Chemistry Connection, Acid Rain, TE** Students research acid rain and determine if it is a problem in their area. **(ADVANCED)**

__ **Teaching Transparency 76, The Water Cycle with Worksheet** Use this Teaching Transparency and the accompanying worksheet to support your classroom discussion of the water cycle.

__ **Quick Lab, Modeling the Water Cycle, SE** Students conduct an experiment to study evaporation and condensation. **(GENERAL)**

__ **Datasheet for Quick Lab, Modeling the Water Cycle, CRF** Students use the datasheet to complete the Quick Lab. **(GENERAL)**

__ **Inquiry Lab, Eutrophication, CRF** Students study eutrophication in lakes and rivers and predict which body of water is affected more by cultural eutrophication. **(GENERAL)**

__ **Skills Practice Lab, Stream Quality Monitoring, CRF** Students collect and identify stream microinvertebrates to assess the biological health of a stream. **(GENERAL)**

__ **Interactive Tutor CD-ROM, Ice, Water, and Vapor** Students practice what they have learned using this interactive activity. **(GENERAL)**

__ **Interactive Tutor CD-ROM, The Water Cycle** Students practice what they have learned using this interactive activity. **(GENERAL)**

CLOSE *(10 minutes)*

KEY **SE** = Student Edition **TE** = Teacher Edition **CRF** = Chapter Resource File

Lesson Plan

__ **Section Review, SE** Students answer end-of-section vocabulary, key ideas, critical thinking, and concept mapping questions. **(GENERAL)**

__ **Reteaching, Water Budget Diagram, TE** Students look at regional photos to review the water cycle and discuss balanced water budgets. **(BASIC)**

__ **Quiz, TE** Students answer 2 questions about this section. **(GENERAL)**

__ **Section Quiz, CRF** Use this quiz to assess students' understanding of the section. **(GENERAL) Also in Spanish**

__ **Spanish Section Quiz, Spanish Assessments** Use this quiz to assess students' understanding of the section. **(GENERAL)**

__ **Alternative Assessment, Skit, TE** Students work in groups to act out key processes in the water cycle. **(GENERAL)**

KEY **SE** = Student Edition **TE** = Teacher Edition **CRF** = Chapter Resource File

Lesson Plan

Section 2 Stream Erosion

PACING

Regular Schedule:	**with lab(s): 2 days**	**without lab(s): 1 day**
Block Schedule:	**with lab(s): 1 day**	**without lab(s): 0.5 day**

OBJECTIVES

1. **Summarize** how a river develops.
2. **Describe** the parts of a river system.
3. **Explain** factors that affect the erosive ability of a river.
4. **Describe** how erosive factors affect the evolution of a river channel.

NATIONAL SCIENCE EDUCATION STANDARDS

PS 2a: Atoms interact with one another by transferring or sharing electrons that are furthest from the nucleus. These outer electrons govern the chemical properties of the element.

UCP 4: Evolution is a series of changes, some gradual and some sporadic, that accounts for the present form and function of objects, organisms, and natural and designed systems. The general idea of evolution is that the present arises from materials and forms of the past. Although evolution is most commonly associated with the biological theory explaining the process of descent with modification of organisms from common ancestors, evolution also describes changes in the universe.

FOCUS *(5 minutes)*

__ **Overview, TE** Review the objectives listed in the Student Edition. **(GENERAL)**

__ **PowerPoint® Resources** Use the customizable presentation to enhance your lecture.

__ **Bellringer, TE** Ask students to make a list of words that describe rivers. **(GENERAL)**

__ **Bellringer Transparency** Use this transparency as students enter the classroom and find their seats.

MOTIVATE *(10 minutes)*

__ **Internet Activity, Watersheds, TE** Students identify their watershed and research ways to protect it. **(GENERAL)**

__ **Internet Activity, Watersheds, CRF** This worksheet directs student research for the topic presented in the Teacher Edition. **(GENERAL)**

KEY **SE** = Student Edition **TE** = Teacher Edition **CRF** = Chapter Resource File

Lesson Plan

TEACH *(65 minutes)*

__ **Directed Reading, CRF** These worksheets reinforce basic concepts and vocabulary presented in the section. **(BASIC)**

__ **Group Activity, Field Trip, TE** Students visit a stream or river in their watershed and record down observations. **(GENERAL)**

__ **Teaching Transparency 77, Stream Gradient and Channel Erosion with Worksheet** Use this Teaching Transparency and the accompanying worksheet to support your classroom discussion of stream gradient and channel erosion.

__ **The Arts Connection, The Big Muddy, TE** Students discuss why the Mississippi River plays such an important role in American life and culture. **(ADVANCED)**

__ **Skill Builder, Vocabulary, TE** Students review the origins of the words *tributary* and *meander.* **(BASIC)**

__ **Graphic Organizer, Cause-and-Effect Map, SE** Use the graphic organizer to help students develop their analytical skills. **(GENERAL)**

__ **Inquiry Lab, Sediments and Water, SE** Students determine the erosive effect of water on various types of sediment. **(GENERAL)**

__ **Datasheet for Chapter Lab, Inquiry Lab, Sediments and Water, CRF** Students use the datasheet to complete the Chapter Lab. **(GENERAL)**

__ **Maps in Action, World Watershed Sediment Yield, SE** Students enhance their map-reading skills by answering questions about a map showing the world's watersheds. **(GENERAL)**

__ **Teaching Transparency with Worksheet, World Watershed Sediment Yield** Use this Teaching Transparency and the accompanying worksheet to help students perform the Maps in Action activity.

__ **Discussion, Sediment Load and Land Use, TE** Students discuss why areas in Southeast Asia have relatively high sediment yields. **(GENERAL)**

__ **Mapping Expeditions, What Comes Down Must Go. . .Where?** Students observe where rainwater goes after it rains and identify sources of pollution. **(GENERAL)**

__ **Interactive Tutor CD-ROM, Physical and Chemical Weathering** Students practice what they have learned using this interactive activity. **(GENERAL)**

CLOSE *(10 minutes)*

__ **Section Review, SE** Students answer end-of-section vocabulary, key ideas, critical thinking, and concept mapping questions. **(GENERAL)**

__ **Reteaching, Flowchart, TE** Students create flowcharts showing the stages of evolution of streams. **(BASIC)**

__ **Quiz, TE** Students answer 3 questions about this section. **(GENERAL)**

__ **Section Quiz, CRF** Use this quiz to assess students' understanding of the section. **(GENERAL) Also in Spanish**

KEY **SE** = Student Edition **TE** = Teacher Edition **CRF** = Chapter Resource File

__ **Spanish Section Quiz, Spanish Assessments** Use this quiz to assess students' understanding of the section. **(GENERAL)**

__ **Alternative Assessment, Build a Watershed, TE** Students build models of a watershed using sand, clay, and a plastic container. **(GENERAL)**

KEY **SE** = Student Edition **TE** = Teacher Edition **CRF** = Chapter Resource File

Section 3 Stream Deposition

PACING

Regular Schedule:	**with lab(s): N/A**	**without lab(s): 2 days**
Block Schedule:	**with lab(s): N/A**	**without lab(s): 1 day**

OBJECTIVES

1. **Explain** the two types of stream deposition.
2. **Describe** one advantage and one disadvantage of living in a floodplain.
3. **Identify** three methods of flood control.
4. **Describe** the life cycle of a lake.

NATIONAL SCIENCE EDUCATION STANDARDS

PS 2a: Atoms interact with one another by transferring or sharing electrons that are furthest from the nucleus. These outer electrons govern the chemical properties of the element.

UCP 4: Evolution is a series of changes, some gradual and some sporadic, that accounts for the present form and function of objects, organisms, and natural and designed systems. The general idea of evolution is that the present arises from materials and forms of the past. Although evolution is most commonly associated with the biological theory explaining the process of descent with modification of organisms from common ancestors, evolution also describes changes in the universe.

FOCUS *(5 minutes)*

__ **Overview, TE** Review the objectives listed in the Student Edition. **(GENERAL)**

__ **PowerPoint® Resources** Use the customizable presentation to enhance your lecture.

__ **Bellringer, TE** Ask students to write a short paragraph that describes what they think happens to all of the sediment eroded and carried by a river. **(GENERAL)**

__ **Bellringer Transparency** Use this transparency as students enter the classroom and find their seats.

MOTIVATE *(10 minutes)*

__ **Demonstration, Changing Land Use, TE** Students investigate how land development affects runoff. **(GENERAL)**

KEY SE = Student Edition **TE** = Teacher Edition **CRF** = Chapter Resource File

Lesson Plan

TEACH *(20 minutes)*

__ **Directed Reading, CRF** These worksheets reinforce basic concepts and vocabulary presented in the section. **(BASIC)**

__ **Debate, To Stay or Not to Stay, TE** Students debate this issue: "Should communities be allowed to rebuild in an area prone to repeated and severe flooding?" **(ADVANCED)**

__ **Skill Builder, Writing, TE** Students research and write an article about an historic U.S. flood. **(ADVANCED)**

__ **Reading Skill Builder, Reading Organizer, TE** Students are encouraged to take notes while reading this section. **(BASIC)**

__ **History Connection, Cradles of Civilization, TE** Students research one of the four "cradles of civilization" and create posters explaining how major waterways influenced its development. **(ADVANCED)**

__ **Quick Lab, Soil Erosion, SE** Students conduct an experiment to study water flow. **(GENERAL)**

__ **Datasheet for Quick Lab, Soil Erosion, CRF** Students use the datasheet to complete the Quick Lab. **(GENERAL)**

__ **Eye on the Environment, The Three Gorges Dam, SE** The benefits and drawbacks of the Three Gorges Dam project are discussed.

__ **Internet Activity, Restoring Rivers, TE** Students research how a dam's removal affects a river and its watershed. **(GENERAL)**

__ **Internet Activity, Restoring Rivers, CRF** This worksheet directs student research for the topic presented in the Teacher Edition. **(GENERAL)**

CLOSE *(10 minutes)*

__ **Section Review, SE** Students answer end-of-section vocabulary, key ideas, critical thinking, and concept mapping questions. **(GENERAL)**

__ **Reteaching, Peer Reviews, TE** Students write questions about this section and quiz each other. **(BASIC)**

__ **Quiz, TE** Students answer 2 questions about this section. **(GENERAL)**

__ **Section Quiz, CRF** Use this quiz to assess students' understanding of the section. **(GENERAL) Also in Spanish**

__ **Spanish Section Quiz, Spanish Assessments** Use this quiz to assess students' understanding of the section. **(GENERAL)**

__ **Alternative Assessment, Sediment Stories, TE** Students create collages illustrating how and where rivers deposit their sediment loads. **(BASIC)**

KEY **SE** = Student Edition **TE** = Teacher Edition **CRF** = Chapter Resource File

Lesson Plan

End of Chapter Review and Assessment

PACING

Regular Schedule: **with lab(s): N/A** **without lab(s): 2 days**
Block Schedule: **with lab(s): N/A** **without lab(s): 1 day**

__ **Chapter Review, SE** Students answer end-of-chapter vocabulary, key ideas, critical thinking, math, writing, and interpreting graphics questions. **(GENERAL)**

__ **Alternative Assessment, Biography of a River, TE** Students research the history and development of a river and write its biography. **(ADVANCED)**

__ **Standardized Test Prep, SE** Have students complete the standardized test prep to help them prepare for standardized tests. **(GENERAL)**

__ **Holt PuzzlePro®** Use this software to create crossword puzzles and word searches that make learning vocabulary fun. **(GENERAL)**

__ **Concept Review, CRF** Use this worksheet to assess student's understanding of the main concepts presented in the chapter. **(GENERAL)**

__ **Critical Thinking, CRF** Students use this worksheet to develop and practice critical thinking skills. **(ADVANCED)**

__ **Math Skills, CRF** This worksheet helps reinforce students' understanding of scientific notation. **(GENERAL)**

__ **Graphing Skills, CRF** This worksheet helps reinforce students' understanding of pie graphs. **(GENERAL)**

__ **Concept Review, Study Guide, River Systems** Use this worksheet to assess student's understanding of the main concepts presented in the chapter. **(GENERAL)**

__ **Concept Review, Spanish Study Guide, River Systems** This worksheet helps reinforce English-language learners' understanding of the main concepts presented in the chapter. **(GENERAL)**

__ **Chapter Test A, CRF** Assign questions from the appropriate test for chapter assessment. **(GENERAL)**

__ **Chapter Test B, CRF** Assign questions from the appropriate test for chapter assessment. **(ADVANCED)**

__ **Spanish Chapter Test A, Spanish Assessments** Assign questions from the appropriate test for chapter assessment. **(GENERAL)**

__ **Test Generator, One-Stop Planner** Create a customized homework assignment, quiz, or test using the HRW Test Generator program. **(GENERAL)**

__ **Test Item Listing, CRF** Use the Test Item Listing to identify questions for a customized homework, quiz, or test.

__ **PowerPoint® Resources** Use the customizable presentation to help students practice standardized test-taking skills.

KEY **SE** = Student Edition **TE** = Teacher Edition **CRF** = Chapter Resource File

OTHER RESOURCE OPTIONS

__ **HRW Earth Science Video, Segment 10, "Water and Erosion"** This segment examines the subject of rivers as agents of erosion. The transport of sediment is followed from the source of a river to deposition in the ocean, where with time and deep burial the sediment solidifies and becomes rock. **(GENERAL)**

__ **Brain Food Video Quiz** Use this video resource to assess student's progress. **(GENERAL)**

KEY **SE** = Student Edition **TE** = Teacher Edition **CRF** = Chapter Resource File

Chapter 16 Groundwater

Chapter Opener

__ **Pre-Reading Activity, Key-Term Fold, SE**

__ **Using the Figure, Warmth in a Cold Climate, TE**

__ **Long-Term Projects, Precipitation and the Water Table, Long-Term Projects**
Students discover how the level of the water table varies with the amount of precipitation. **(ADVANCED)**

Section 1 Water Beneath the Surface

PACING

Regular Schedule:	**with lab(s): 2 days**	**without lab(s): 1 day**
Block Schedule:	**with lab(s): 1 day**	**without lab(s): 0.5 day**

OBJECTIVES

1. **Identity** properties of aquifers that affect the flow of groundwater.
2. **Describe** the water table and its relationship to the land surface.
3. **Compare** wells, springs, and artesian formations.
4. **Describe** two land features formed by hot groundwater.

NATIONAL SCIENCE EDUCATION STANDARDS

LS 4e: Human beings live within the world's ecosystems. Increasingly, humans modify ecosystems as a result of population growth, technology, and consumption. Human destruction of habitats through direct harvesting, pollution, atmospheric changes, and other factors is threatening current global stability, and if not addressed, ecosystems will be irreversibly affected.

SPSP 6b: Understanding basic concepts and principles of science and technology should precede active debate about the economics, policies, politics, and ethics of various science- and technology-related challenges. However, understanding science alone will not resolve local, national, or global challenges.

UCP 1: The natural and designed world is complex; it is too large and complicated to investigate and comprehend all at once. Scientists and students learn to define small portions for the convenience of investigation. The units of investigation can be referred to as "systems." A system is an organized group of related objects or components that form a whole. Systems can consist, for example, of organisms, machines, fundamental particles, galaxies, ideas, numbers, transportation, and education. Systems have boundaries, components, resources flow (input and output), and feedback.

KEY **SE** = Student Edition **TE** = Teacher Edition **CRF** = Chapter Resource File

Lesson Plan

FOCUS *(5 minutes)*

__ **Overview, TE** Review the objectives listed in the Student Edition. **(GENERAL)**

__ **PowerPoint® Resources** Use the customizable presentation to enhance your lecture.

__ **Bellringer, TE** Students are asked where the water that flows from the faucets in their homes comes from. **(GENERAL)**

__ **Bellringer Transparency** Use this transparency as students enter the classroom and find their seats.

MOTIVATE *(10 minutes)*

__ **Using the Figure, Compare, TE** Students use the figure showing differences in porosity to describe the three types of sediment. **(GENERAL)**

TEACH *(65 minutes)*

__ **Directed Reading, CRF** These worksheets reinforce basic concepts and vocabulary presented in the section. **(BASIC)**

__ **Teaching Transparency 79, Porosity and Permeability with Worksheet** Use this Teaching Transparency and the accompanying worksheet to support your classroom discussion of porosity and permeability.

__ **Using the Figure, Replenishing Groundwater, TE** Students study the figure of a porous and a permeable rock and describe the differences between them. **(GENERAL)**

__ **Quick Lab, Permeability, SE** Students compare the drainage rates and permeability of sand and gravel. **(GENERAL)**

__ **Datasheet for Quick Lab, Permeability, CRF** Students use the datasheet to complete the Quick Lab. **(GENERAL)**

__ **Demonstration, Comprehension Check, TE** Students demonstrate the concepts of permeability and impermeability. **(GENERAL)**

__ **Using the Figure, Extreme Conditions, TE** Using the figure illustrating groundwater zones, students identify and describe each layer. **(GENERAL)**

__ **Group Activity, Model an Aquifer, TE** Students create models of an aquifer. **(GENERAL)**

__ **Reading Skill Builder, Paired Summarizing, TE** Students read and summarize the zones of aquifers. **(BASIC)**

__ **Teaching Transparency 79, Zones of Aquifers with Worksheet** Use this Teaching Transparency and the accompanying worksheet to support your classroom discussion of zones of aquifers.

__ **Teaching Transparency 80, Topography and the Water Table with Worksheet** Use this Teaching Transparency and the accompanying worksheet to support your classroom discussion of topography and the water table.

KEY **SE** = Student Edition **TE** = Teacher Edition **CRF** = Chapter Resource File

__ **Using the Figure, Going Down the Drain, TE** Students identify the location of the drain shown in the photo. **(BASIC)**

__ **Graphic Organizer, Venn Diagram, SE** Use the Graphic Organizer to help students develop their analytical skills. **(GENERAL)**

__ **Discussion, Comprehension, TE** Students discuss the role of caprock in an artesian well or spring. **(BASIC)**

__ **Using the Figure, Slope, TE** Students study the diagram of an artesian formation and speculate about why the slope enhances the formation. **(ADVANCED)**

__ **Skill Builder, Vocabulary, TE** Students learn the history of the word *artesian.* **(GENERAL)**

__ **Skills Practice Lab, Porosity, SE** Students measure and compare the porosity of beads. **(GENERAL)**

__ **Datasheet for Chapter Lab, Skills Practice Lab, Porosity, CRF** Students use the datasheet to complete the Chapter Lab. **(GENERAL)**

__ **Maps in Action, Water Level in the Southern Ogallala, SE** Students enhance their map-reading skills by answering questions about a graphic map showing the water level in the southern Ogallala. **(GENERAL)**

__ **Teaching Transparency 81, Water Level in the Southern Ogallala with Worksheet** Use this Teaching Transparency and the accompanying worksheet to help students perform the Maps in Action activity.

__ **Mapping Expeditions, What Comes Down Must Go…Where? SE** Students observe where rainwater goes after it rains and they identify sources of pollution. **(GENERAL)**

__ **Making Models Lab, Making a Classroom Geyser, CRF** Students make models showing how a geyser functions and predict when the geyser will erupt. **(GENERAL)**

CLOSE *(10 minutes)*

__ **Section Review** Students answer end of section vocabulary, key ideas, critical thinking, and concept mapping questions **(GENERAL)**

__ **Homework, Home Water Use, TE** Students list appliances in their homes that use water and then check which ones are designed to use "less" water. **(GENERAL)**

__ **Reteaching, Diagrams, TE** Students make diagrams of formations that hold groundwater. **(BASIC)**

__ **Quiz, TE** Students answer 3 questions about this section. **(GENERAL)**

__ **Section Quiz, CRF** Use this quiz to assess students' understanding of the section. **(GENERAL) Also in Spanish**

__ **Spanish Section Quiz, Spanish Assessments** Use this quiz to assess students' understanding of the section. **(GENERAL)**

__ **Alternative Assessment, Assessing Groundwater, TE** Students list criteria to study when preparing a report about groundwater resources. **(GENERAL)**

KEY **SE** = Student Edition **TE** = Teacher Edition **CRF** = Chapter Resource File

Lesson Plan

Section 2 Groundwater and Chemical Weathering

PACING

Regular Schedule: **with lab(s): N/A** **without lab(s): 2 days**
Block Schedule: **with lab(s): N/A** **without lab(s): 1 day**

OBJECTIVES

1. **Describe** how water chemically weathers rock.
2. **Explain** how caverns and sinkholes form.
3. **Identify** two features of karst topography.

NATIONAL SCIENCE EDUCATION STANDARDS

PS 3c: A large number of important reactions involve the transfer of either electrons (oxidation/reduction reactions) or hydrogen ions (acid/base reactions) between reacting ions, molecules, or atoms. In other reactions, chemical bonds are broken by heat or light to form very reactive radicals with electrons ready to form new bonds. Radical reactions control many processes such as the presence of ozone and greenhouse gases in the atmosphere, burning and processing of fossil fuels, the formation of polymers, and explosions.

UCP 3: Although most things are in the process of becoming different–changing–some properties of objects and processes are characterized by constancy, including the speed of light, the charge of an electron, and the total mass plus energy in the universe. Changes might occur, for example, in properties of materials, position of objects, motion, and form and function of systems. Interactions within and among systems result in change. Changes vary in rate, scale, and pattern, including trends and cycles.

UCP 4: Evolution is a series of changes, some gradual and some sporadic, that accounts for the present form and function of objects, organisms, and natural and designed systems. The general idea of evolution is that the present arises from materials and forms of the past. Although evolution is most commonly associated with the biological theory explaining the process of descent with modification of organisms from common ancestors, evolution also describes changes in the universe.

FOCUS *(5 minutes)*

__ **Overview, TE** Review the objectives listed in the Student Edition. **(GENERAL)**

__ **PowerPoint® Resources** Use the customizable presentation to enhance your lecture.

__ **Bellringer, TE** Students are asked to describe how minerals in water can affect objects that are in extensive contact with the water. **(GENERAL)**

__ **Bellringer Transparency** Use this transparency as students enter the classroom and find their seats.

KEY **SE** = Student Edition **TE** = Teacher Edition **CRF** = Chapter Resource File

Lesson Plan

MOTIVATE *(25 minutes)*

__ **Quick Lab, Chemical Weathering, SE** Students observe the effects of chemical weathering on several different kinds of rock. **(GENERAL)**

__ **Datasheet for Quick Lab, Chemical Weathering, CRF** Students use the datasheet to complete the Quick Lab. **(GENERAL)**

TEACH *(20 minutes)*

__ **Directed Reading, CRF** These worksheets reinforce basic concepts and vocabulary presented in the section. **(BASIC)**

__ **Using the Figure, Cave Formations, TE** Students use the photo of a cavern to discuss some unusual cave formations. **(GENERAL)**

__ **Chemistry Connection, Rock and Soil, TE** Students research the connection between the type of soil found in their area and bedrock. **(ADVANCED)**

__ **Reading Skill Builder, Reading Hint, TE** Students learn a mnemonic for the words *stalagmite* and *stalactite*. **(BASIC)**

__ **Biology Connection, Cave Vision, TE** Students discuss why many animals that live in caves are blind. **(ADVANCED)**

__ **Skill Builder, Writing, TE** Students write a letter to their local newspaper explaining how they feel about a sinkhole that "swallowed" several cars. **(ADVANCED)**

__ **Impact on Society, Disappearing Land, SE** Causes and effects of the Winter Park, Florida sinkhole are discussed.

__ **Inquiry Lab, Cave Formations and Ecology, CRF** Students observe how cave formations grow and identify ways in which that growth can be changed. **(GENERAL)**

__ **Interactive Tutor CD-ROM, Physical and Chemical Weathering** Students practice what they have learned using this interactive activity. **(GENERAL)**

CLOSE *(10 minutes)*

__ **Section Review, SE** Students answer end-of-section vocabulary, key ideas, critical thinking, and concept mapping questions **(GENERAL)**

__ **Reteaching, Illustrating Cave Formations, TE** Students illustrate how water flows through rock to create caves and cave formations. **(BASIC)**

__ **Quiz, TE** Students answer 2 questions about this section. **(GENERAL)**

__ **Section Quiz, CRF** Use this quiz to assess students' understanding of the section. **(GENERAL) Also in Spanish**

__ **Spanish Section Quiz, Spanish Assessments** Use this quiz to assess students' understanding of the section. **(GENERAL)**

__ **Alternative Assessment, Guided Tour, TE** Students imagine that they are cavern guides and create brochures describing a cavern's history and features. **(ADVANCED)**

KEY **SE** = Student Edition **TE** = Teacher Edition **CRF** = Chapter Resource File

Lesson Plan

End of Chapter Review and Assessment

PACING

Regular Schedule: **with lab(s): N/A** **without lab(s): 2 days**
Block Schedule: **with lab(s): N/A** **without lab(s): 1 day**

__ **Chapter Review, SE** Students answer end-of-chapter vocabulary, key ideas, critical thinking, math, writing, and interpreting graphics questions. **(GENERAL)**

__ **Alternative Assessment, The Water Beneath Your Feet, TE** In groups, students research the groundwater resources and characteristics in their region or state. **(GENERAL)**

__ **Standardized Test Prep, SE** Have students complete the standardized test prep to help them prepare for standardized tests. **(GENERAL)**

__ **Holt PuzzlePro®** Use this software to create crossword puzzles and word searches that make learning vocabulary fun. **(GENERAL)**

__ **Section Quiz, CRF** Use this quiz to assess students' understanding of the section. **(GENERAL) Also in Spanish**

__ **Concept Review, CRF** Use this worksheet to assess student's understanding of the main concepts presented in the chapter. **(GENERAL)**

__ **Critical Thinking, CRF** Students use critical thinking skills to answer questions about groundwater. **(GENERAL)**

__ **Math Skills, CRF** This worksheet helps reinforce students' understanding of Geometry. **(GENERAL)**

__ **Graphing Skills, CRF** This worksheet helps reinforce students' understanding of Bar Graphs. **(GENERAL)**

__ **Concept Review, Study Guide, Plate Tectonics** Use this worksheet to assess student's understanding of the main concepts presented in the chapter. **(GENERAL)**

__ **Concept Review, Spanish Study Guide, Plate Tectonics** This worksheet helps reinforce English-language learners' understanding of the main concepts presented in the chapter. **(GENERAL)**

__ **Chapter Test A, CRF** Assign questions from the appropriate test for chapter assessment. **(GENERAL)**

__ **Chapter Test B, CRF** Assign questions from the appropriate test for chapter assessment. **(ADVANCED)**

__ **Spanish Chapter Test A, Spanish Assessments** Assign questions from the appropriate test for chapter assessment. **(GENERAL)**

__ **Test Generator, One-Stop Planner** Create a customized homework assignment, quiz, or test using the HRW Test Generator program. **(GENERAL)**

__ **Test Item Listing, CRF** Use the Test Item Listing to identify questions for a customized homework, quiz, or test.

__ **PowerPoint® Resources** Use the customizable presentation to help students practice standardized test-taking skills.

KEY **SE** = Student Edition **TE** = Teacher Edition **CRF** = Chapter Resource File

OTHER RESOURCE OPTIONS

__ **Brain Food Video Quiz** Use this video resource to assess student's progress. **(GENERAL)**

KEY **SE** = Student Edition **TE** = Teacher Edition **CRF** = Chapter Resource File

Chapter 17 Glaciers

Chapter Opener

__ **Pre-Reading Activity, Double Door, SE**

__ **Using the Figure, Blue and White Ice, TE (ADVANCED)**

Section 1 Glaciers: Moving Ice

PACING

Regular Schedule:	**with lab(s): N/A**	**without lab(s): 2 days**
Block Schedule:	**with lab(s): N/A**	**without lab(s): 1 day**

OBJECTIVES

1. **Describe** how glaciers form.
2. **Compare** two main kinds of glaciers.
3. **Explain** two processes by which glaciers move.
4. **Describe** three features of glaciers.

NATIONAL SCIENCE EDUCATION STANDARDS

UCP 4: Evolution is a series of changes, some gradual and some sporadic, that accounts for the present form and function of objects, organisms, and natural and designed systems. The general idea of evolution is that the present arises from materials and forms of the past. Although evolution is most commonly associated with the biological theory explaining the process of descent with modification of organisms from common ancestors, evolution also describes changes in the universe.

SPSP 5c: Some hazards, such as earthquakes, volcanic eruptions, and severe weather, are rapid and spectacular. But there are slow and progressive changes that also result in problems for individuals and societies. For example, change in stream channel position, erosion of bridge foundations, sedimentation in lakes and harbors, coastal erosions, and continuing erosion and wasting of soil and landscapes can all negatively affect society.

FOCUS *(5 minutes)*

__ **Overview, TE** Review the objectives listed in the Student Edition. **(GENERAL)**

__ **PowerPoint® Resources** Use the customizable presentation to enhance your lecture.

__ **Bellringer, TE** Students are asked what they know about ice and glaciers. **(GENERAL)**

__ **Bellringer Transparency** Use this transparency as students enter the classroom and find their seats.

KEY **SE** = Student Edition **TE** = Teacher Edition **CRF** = Chapter Resource File

Lesson Plan

MOTIVATE *(10 minutes)*

__ **Activity, Modeling Glacial Ice Formation, TE** Students make paper snowflakes and then crush the snowflakes to represent firn and glacial ice. **(GENERAL)**

TEACH *(20 minutes)*

__ **Directed Reading, CRF** These worksheets reinforce basic concepts and vocabulary presented in the section. **(BASIC)**

__ **Internet Activity, Glacial Sounds, TE** Students predict what sounds glaciers make and then go online to listen to the actual sounds of glaciers. **(GENERAL)**

__ **Internet Activity, Glacial Sounds, CRF** This worksheet directs student research for the topic presented in the Teacher Edition. **(GENERAL)**

__ **Quick Lab, Slipping Ice, SE** Students learn about glacier movement by squeezing shaved ice. **(GENERAL)**

__ **Datasheet for Quick Lab, Slipping Ice, CRF** Students use the datasheet to complete the Quick Lab. **(GENERAL)**

__ **Teaching Transparency 83, Internal Plastic Flow with Worksheet** Use this Teaching Transparency and the accompanying worksheet to support your classroom discussion of internal plastic flow.

__ **Maps in Action, Gulkana Glacier, SE** Students enhance their map-reading skills by answering questions about a graphic map showing the Gulkana Glacier. **(GENERAL)**

__ **Teaching Transparency 88, Gulkana Glacier with Worksheet** Use this teaching transparency and the accompanying worksheet to help students perform the Maps in Action activity.

__ **Internet Activity, Photographic Evidence, TE** Students research the Gulkana Glacier and write an essay describing the Gulkana landscape. **(GENERAL)**

__ **Internet Activity, Photographic Evidence, CRF** This worksheet directs student research for the topic presented in the Teacher Edition. **(GENERAL)**

CLOSE *(10 minutes)*

__ **Section Review, SE** Students answer end-of-section vocabulary, key ideas, critical thinking, and concept mapping questions. **(GENERAL)**

__ **Reteaching, Glacial Movement, TE** In pairs, students review both types of glacial movement. **(BASIC)**

__ **Quiz, TE** Students answer 3 questions about this section. **(GENERAL)**

__ **Section Quiz, CRF** Use this quiz to assess students' understanding of the section. **(GENERAL) Also in Spanish**

__ **Spanish Section Quiz, Spanish Assessments** Use this quiz to assess students' understanding of the section. **(GENERAL)**

KEY **SE** = Student Edition **TE** = Teacher Edition **CRF** = Chapter Resource File

Lesson Plan

__ **Alternative Assessment, Glacial Variety and Features, TE** Students create slide presentations showing images of different glacial types and features. **(GENERAL)**

KEY SE = Student Edition **TE** = Teacher Edition **CRF** = Chapter Resource File

Section 2 Glacial Erosion and Deposition

PACING

Regular Schedule: **with lab(s): N/A** **without lab(s): 2 days**
Block Schedule: **with lab(s): N/A** **without lab(s): 1 day**

OBJECTIVES

1. **Describe** the landscape features that are produced by glacial erosion.
2. **Name** and describe five features formed by glacial deposition.
3. **Explain** how glacial lakes form.

NATIONAL SCIENCE EDUCATION STANDARDS

UCP 4: Evolution is a series of changes, some gradual and some sporadic, that accounts for the present form and function of objects, organisms, and natural and designed systems. The general idea of evolution is that the present arises from materials and forms of the past. Although evolution is most commonly associated with the biological theory explaining the process of descent with modification of organisms from common ancestors, evolution also describes changes in the universe.

SPSP 5c: Some hazards, such as earthquakes, volcanic eruptions, and severe weather, are rapid and spectacular. But there are slow and progressive changes that also result in problems for individuals and societies. For example, change in stream channel position, erosion of bridge foundations, sedimentation in lakes and harbors, coastal erosions, and continuing erosion and wasting of soil and landscapes can all negatively affect society.

FOCUS *(5 minutes)*

__ **Overview, TE** Review the objectives listed in the Student Edition. **(GENERAL)**

__ **PowerPoint® Resources** Use the customizable presentation to enhance your lecture.

__ **Bellringer, TE** Students are asked to describe how glaciers behave like a steam shovel, a chisel, sandpaper, and a conveyor belt. **(GENERAL)**

__ **Bellringer Transparency** Use this transparency as students enter the classroom and find their seats.

MOTIVATE *(10 minutes)*

__ **Demonstration, Glacial Erosion and Deposition Simulation, TE** Students use salt to simulate how glacial movement affects land. **(GENERAL)**

TEACH *(20 minutes)*

__ **Directed Reading, CRF** These worksheets reinforce basic concepts and vocabulary presented in the section. **(BASIC)**

KEY **SE** = Student Edition **TE** = Teacher Edition **CRF** = Chapter Resource File

__ **Reading Skill Builder, Reading Organizer, TE** Students are encouraged to take notes while reading this section. **(BASIC)**

__ **Skill Builder, Vocabulary, TE** Students review the meaning of the word *arête*. **(GENERAL)**

__ **Using the Figure, Comprehension Check, TE** Students review landforms created by glaciers to learn about the path and action of glaciers. **(GENERAL)**

__ **Teaching Transparency 84, Landforms Created by Glacial Erosion with Worksheet** Use this Teaching Transparency and the accompanying worksheet to support your classroom discussion of landforms created by glacial erosion.

__ **Quick Lab, Glacial Erosion, SE** Students perform an experiment to learn about glacier erosion. **(GENERAL)**

__ **Datasheet for Quick Lab, Glacial Erosion, CRF** Students use the datasheet to complete the Quick Lab. **(GENERAL)**

__ **Activity, Glacial Erosion Theories, TE** Students research the development of glacial erosion hypotheses and write essays detailing their findings. **(ADVANCED)**

__ **Discussion, Comprehension Check, TE** Students imagine that they have discovered a large boulder and identify the glacial deposit. **(GENERAL)**

__ **Teaching Transparency 85, Features of Glacial Deposition with Worksheet** Use this Teaching Transparency and the accompanying worksheet to support your classroom discussion of features of glacial deposition.

__ **Graphic Organizer, Spider Map, SE** Use the graphic organizer to help students develop their analytical skills. **(GENERAL)**

__ **Using the Figure, Describing Glacial Landforms, TE** Students use the figure of glacial landforms to describe their features and how they formed. **(GENERAL)**

__ **Activity, Effects of Glacial Erosion and Deposition, TE** Students create a two-column table to review which process yield the various landforms. **(BASIC)**

__ **Internet Activity, More Glacial Landforms and Features, TE** Students research topographical features formed by glacial deposition. **(GENERAL)**

__ **Internet Activity, More Glacial Landforms and Features, CRF** This worksheet directs student research for the topic presented in the Teacher Edition. **(GENERAL)**

__ **Geography Connection, Local Glacial Formations, TE** In pairs, students research the glacial history of the continents. **(ADVANCED)**

__ **Group Activity, Types of Glacial Lakes, TE** Students work in groups to research each of the three types of glacial lakes. **(GENERAL)**

__ **Biology Connection, Pink Salt Lakes, TE** Students research why Lake Owens is pink. **(GENERAL)**

__ **Group Activity, Skit, TE** Students perform skits based on their research of the floods that formed the Channeled Scablands. **(ADVANCED)**

__ **Inquiry Lab, Glacial Deposition, CRF** Students use sand, wood, and ice blocks to model till deposits and moraine formation by receding glaciers. **(GENERAL)**

KEY **SE** = Student Edition **TE** = Teacher Edition **CRF** = Chapter Resource File

__ **Making Models Lab, Melted Glacier Formations, CRF** Students model stratified drift, eskers, and kettles to study how melting ice shapes land features. **(GENERAL)**

CLOSE *(10 minutes)*

__ **Section Review, SE** Students answer end-of-section vocabulary, key ideas, critical thinking, and concept mapping questions. **(GENERAL)**

__ **Homework, Glacial Deposition and Erosion Gallery, TE** Students create scrapbooks showing the different effects of glacial erosion. **(GENERAL)**

__ **Homework, Fiord Formation, TE** Students research the discovery of the Harriman Fiord in Alaska. **(BASIC)**

__ **Reteaching, Chart Glacial Landforms, TE** In groups, students review the types of glacial landforms, their descriptions, and their formation. **(BASIC)**

__ **Quiz, TE** Students answer 3 questions about this section. **(GENERAL)**

__ **Section Quiz, CRF** Use this quiz to assess students' understanding of the section. **(GENERAL) Also in Spanish**

__ **Spanish Section Quiz, Spanish Assessments** Use this quiz to assess students' understanding of the section. **(GENERAL)**

__ **Alternative Assessment, Modeling Glacial Landforms, TE** Students make three-dimensional models showing different landforms created by glaciation. **(ADVANCED)**

KEY **SE** = Student Edition **TE** = Teacher Edition **CRF** = Chapter Resource File

Lesson Plan

Section 3 Ice Ages

PACING

Regular Schedule: **with lab(s): 2 days** **without lab(s): 1 day**
Block Schedule: **with lab(s): 1 day** **without lab(s): 0.5 day**

OBJECTIVES

1. **Describe** glacial and interglacial periods within an ice age.
2. **Summarize** the theory that best accounts for the ice ages.

NATIONAL SCIENCE EDUCATION STANDARDS

UCP 4: Evolution is a series of changes, some gradual and some sporadic, that accounts for the present form and function of objects, organisms, and natural and designed systems. The general idea of evolution is that the present arises from materials and forms of the past. Although evolution is most commonly associated with the biological theory explaining the process of descent with modification of organisms from common ancestors, evolution also describes changes in the universe.

SPSP 5c: Some hazards, such as earthquakes, volcanic eruptions, and severe weather, are rapid and spectacular. But there are slow and progressive changes that also result in problems for individuals and societies. For example, change in stream channel position, erosion of bridge foundations, sedimentation in lakes and harbors, coastal erosions, and continuing erosion and wasting of soil and landscapes can all negatively affect society.

ES 3c: Interactions among the solid earth, the oceans, the atmosphere, and organisms have resulted in the ongoing evolution of the earth system. We can observe some changes such as earthquakes and volcanic eruptions on a human time scale, but many processes such as mountain building and plate movements take place over hundreds of millions of years.

FOCUS *(5 minutes)*

___ **Overview, TE** Review the objectives listed in the Student Edition. **(GENERAL)**

___ **PowerPoint® Resources** Use the customizable presentation to enhance your lecture.

___ **Bellringer, TE** Students are asked to describe what they think Earth would be like during an ice age. **(GENERAL)**

___ **Bellringer Transparency** Use this transparency as students enter the classroom and find their seats.

MOTIVATE *(10 minutes)*

___ **Discussion, Ice Age Mammals, TE** Students discuss possible reasons for the extinction of ice age mammals. **(GENERAL)**

KEY SE = Student Edition **TE** = Teacher Edition **CRF** = Chapter Resource File

Lesson Plan

TEACH *(65 minutes)*

__ **Directed Reading, CRF** These worksheets reinforce basic concepts and vocabulary presented in the section. **(BASIC)**

__ **Teaching Transparency 86, Ice Ages of Earth's History with Worksheet** Use this Teaching Transparency and the accompanying worksheet to support your classroom discussion of the ice ages of Earth's history.

__ **Debate, Global Warming, TE** Students research and debate the effects of global warming on the next glacial period. **(ADVANCED)**

__ **Internet Activity, Vanishing Ice Caps, TE** Students research the significance and implications of Greenland's vanishing ice. **(GENERAL)**

__ **Internet Activity, Vanishing Ice Caps, CRF** This worksheet directs student research for the topic presented in the Teacher Edition. **(GENERAL)**

__ **Teaching Transparency 87, The Milankovitch Theory with Worksheet** Use this Teaching Transparency and the accompanying worksheet to support your classroom discussion of the Milankovitch Theory.

__ **Making Models Lab, Glaciers and Sea Level Change, SE** Students construct models showing what would happen if the Antarctic ice sheet melted. **(GENERAL)**

__ **Datasheet for Chapter Lab, Making Models Lab, Glaciers and Sea Level Change, CRF** Students use the datasheet to complete the Chapter Lab. **(GENERAL)**

__ **Eye on the Environment, The Missoula Floods, SE** The formation of the Channeled Scablands is discussed. **(GENERAL)**

__ **Group Activity, Skit, TE** Students perform skits based on their research of Bretz's evidence for the creation of the Channeled Scablands. **(ADVANCED)**

CLOSE *(10 minutes)*

__ **Section Review, SE** Students answer end-of-section vocabulary, key ideas, critical thinking, and concept mapping questions. **(GENERAL)**

__ **Homework, Evidence of Ice Ages, TE** Students research the evidence that supports the existence of ice ages. **(ADVANCED)**

__ **Reteaching, Reading Organizer, TE** Have students outline the chapter material and provide additional detail for each head and subhead. **(BASIC)**

__ **Quiz, TE** Students answer 3 questions about this section. **(GENERAL)**

__ **Section Quiz, CRF** Use this quiz to assess students' understanding of the section. **(GENERAL) Also in Spanish**

__ **Spanish Section Quiz, Spanish Assessments** Use this quiz to assess students' understanding of the section. **(GENERAL)**

__ **Alternative Assessment, Poster Project, TE** Students create three-paneled posters showing Earth's ice ages, glacial, and interglacial periods. **(GENERAL)**

KEY **SE** = Student Edition **TE** = Teacher Edition **CRF** = Chapter Resource File

Lesson Plan

End of Chapter Review and Assessment

PACING

Regular Schedule: **with lab(s): N/A** **without lab(s): 2 days**
Block Schedule: **with lab(s): N/A** **without lab(s): 1 day**

__ **Chapter Review, SE** Students answer vocabulary, key ideas, critical thinking, math, writing, and interpreting graphics questions. **(GENERAL)**

__ **Alternative Assessment, Glacier Brochure, TE** Students create six-paneled brochures explaining how a glacier forms, moves, what it looks like, and how it affects Earth's surface and climate. **(GENERAL)**

__ **Standardized Test Prep, SE** Have students complete the standardized test prep to help them prepare for standardized tests. **(GENERAL)**

__ **Holt PuzzlePro®** Use this software to create crossword puzzles and word searches that make learning vocabulary fun. **(GENERAL)**

__ **Concept Review, CRF** Use this worksheet to assess student's understanding of the main concepts presented in the chapter. **(GENERAL)**

__ **Critical Thinking, CRF** Students use critical thinking skills to answer questions about glaciers. **(GENERAL)**

__ **Math Skills, CRF** This worksheet helps reinforce students' understanding of Exponents. **(GENERAL)**

__ **Graphing Skills, CRF** This worksheet helps reinforce students' understanding of Line Graphs. **(GENERAL)**

__ **Concept Review, Study Guide, Glaciers** Use this worksheet to assess student's understanding of the main concepts presented in the chapter. **(GENERAL)**

__ **Concept Review, Spanish Study Guide, Glaciers** This worksheet helps reinforce English-language learners' understanding of the main concepts presented in the chapter. **(GENERAL)**

__ **Chapter Test A, CRF** Assign questions from the appropriate test for chapter assessment. **(GENERAL)**

__ **Chapter Test B, CRF** Assign questions from the appropriate test for chapter assessment. **(ADVANCED)**

__ **Spanish Chapter Test A, Spanish Assessments** Assign questions from the appropriate test for chapter assessment. **(GENERAL)**

__ **Test Generator, One-Stop Planner** Create a customized homework assignment, quiz, or test using the HRW Test Generator program. **(GENERAL)**

__ **Test Item Listing, CRF** Use the Test Item Listing to identify questions for a customized homework, quiz, or test.

__ **PowerPoint® Resources** Use the customizable presentation to help students practice standardized test-taking skills.

KEY **SE** = Student Edition **TE** = Teacher Edition **CRF** = Chapter Resource File

OTHER RESOURCE OPTIONS

__ **HRW Earth Science Video, Segment 11, "Glaciers and Erosion"** In this segment two types of glaciers–continental and valley glaciers–are described. In addition, the segment discusses the great Pleistocene ice sheets that covered Northern Europe, Canada, and parts of the United States and discusses some of their effects on terrain.

__ **Brain Food Video Quiz** Use this video resource to assess student's progress. **(GENERAL)**

KEY **SE** = Student Edition **TE** = Teacher Edition **CRF** = Chapter Resource File

Chapter 18 Erosion by Wind and Waves

Chapter Opener

__ **Pre-Reading Activity, Pyramid, SE**

__ **Using the Figure, Rock Formation, TE (GENERAL)**

Section 1 Wind Erosion

PACING

Regular Schedule:	**with lab(s): N/A**	**without lab(s): 2 days**
Block Schedule:	**with lab(s): N/A**	**without lab(s): 1 day**

OBJECTIVES

1. **Describe** two ways that wind erodes land.
2. **Compare** the two types of wind deposits.

NATIONAL SCIENCE EDUCATION STANDARDS

ES 3c: Interactions among the solid earth, the oceans, the atmosphere, and organisms have resulted in the ongoing evolution of the earth system. We can observe some changes such as earthquakes and volcanic eruptions on a human time scale, but many processes such as mountain building and plate movements take place over hundreds of millions of years.

UCP 3: Although most things are in the process of becoming different–changing–some properties of objects and processes are characterized by constancy, including the speed of light, the charge of an electron, and the total mass plus energy in the universe. Changes might occur, for example, in properties of materials, position of objects, motion, and form and function of systems. Interactions within and among systems result in change. Changes vary in rate, scale, and pattern, including trends and cycles.

UCP 4: Evolution is a series of changes, some gradual and some sporadic, that accounts for the present form and function of objects, organisms, and natural and designed systems. The general idea of evolution is that the present arises from materials and forms of the past. Although evolution is most commonly associated with the biological theory explaining the process of descent with modification of organisms from common ancestors, evolution also describes changes in the universe.

FOCUS *(5 minutes)*

__ **Overview, TE** Review the objectives listed in the Student Edition. **(GENERAL)**

__ **PowerPoint® Resources** Use the customizable presentation to enhance your lecture.

__ **Bellringer, TE** Students are asked why something as tiny as sand or dust can erode hard surfaces. **(GENERAL)**

KEY **SE** = Student Edition **TE** = Teacher Edition **CRF** = Chapter Resource File

__ **Bellringer Transparency** Use this transparency as students enter the classroom and find their seats.

MOTIVATE *(10 minutes)*

__ **Demonstration, Watching Sand Jump, TE** Use fine and coarse sand to demonstrate how wind moves sand. **(GENERAL)**

TEACH *(20 minutes)*

__ **Directed Reading, CRF** These worksheets reinforce basic concepts and vocabulary presented in the section. **(BASIC)**

__ **Reading Skill Builder, Paired Summarizing, TE** Students read the text in pairs and take turns summarizing what they have read. **(BASIC)**

__ **Activity, Shaping, TE** Rub an emery board across a pencil to demonstrate how abrasion causes blowing sand. **(GENERAL)**

__ **History Connection, Dust Bowl, TE** Students research the Dust Bowl and present their findings. **(ADVANCED)**

__ **Using the Figure, Dune Formation, TE** Use the figure illustrating dune formation to lead a discussion of how dunes form. **(BASIC)**

__ **Graphic Organizer, Comparison Table, TE** Use the graphic organizer to help students develop their analytical skills. **(GENERAL)**

__ **Teaching Transparency 89, Type of Dunes with Worksheet** Use this Teaching Transparency and the accompanying worksheet to support your classroom discussion of types of dunes.

__ **Internet Activity, Great Lakes Dunes, TE** Students research the history of the sand dunes around the North American Great Lakes. **(ADVANCED)**

__ **Internet Activity, Great Lakes Dunes, CRF** This worksheet directs student research for the topic presented in the Teacher Edition. **(ADVANCED)**

__ **Quick Lab, Dune Migration, SE** Students learn about dune migration by using a box, a hair dryer and sand. **(GENERAL)**

__ **Datasheet for Quick Lab, Dune Migration, CRF** Students use the datasheet to complete the Quick Lab. **(GENERAL)**

__ **Using the Figure, Dune Migration, TE** Students use the photograph of a sand dune to determine which way the wind is blowing. **(BASIC)**

__ **Inquiry Lab, Soil Erosion, CRF** Students test different types of soils and ground cover to find the best combination to control soil erosion. **(GENERAL)**

KEY **SE** = Student Edition **TE** = Teacher Edition **CRF** = Chapter Resource File

Lesson Plan

CLOSE *(10 minutes)*

__ **Section Review, SE** Students answer end-of-section vocabulary, key ideas, critical thinking, and concept mapping questions. **(GENERAL)**

__ **Homework, Soil Conservation, TE** Students research some agricultural methods that have caused erosion. **(BASIC)**

__ **Reteaching, Crossword Puzzle, TE** Students create a crossword puzzle using all of the vocabulary words from this section. **(BASIC)**

__ **Quiz, TE** Students answer 2 questions about this section. **(GENERAL)**

__ **Section Quiz, CRF** Use this quiz to assess students' understanding of the section. **(GENERAL) Also in Spanish**

__ **Spanish Section Quiz, Spanish Assessments** Use this quiz to assess students' understanding of the section. **(GENERAL)**

__ **Alternative Assessment, Landform Description, TE** Students describe the landforms they would see if they were flying over a desert. **(GENERAL)**

KEY **SE** = Student Edition **TE** = Teacher Edition **CRF** = Chapter Resource File

Lesson Plan

Section 2 Wave Erosion

PACING

Regular Schedule:	**with lab(s): 2 days**	**without lab(s): 1 day**
Block Schedule:	**with lab(s): 1 day**	**without lab(s): 0.5 day**

OBJECTIVES

1. **Compare** the formation of six features produced by wave erosion.
2. **Explain** how beaches form.
3. **Describe** the features produced by the movement of sand along a shore.

NATIONAL SCIENCE EDUCATION STANDARDS

ES 3c: Interactions among the solid earth, the oceans, the atmosphere, and organisms have resulted in the ongoing evolution of the earth system. We can observe some changes such as earthquakes and volcanic eruptions on a human time scale, but many processes such as mountain building and plate movements take place over hundreds of millions of years.

SAI 1a: Identify questions and concepts that guide scientific investigations. Students should formulate a testable hypothesis and demonstrate the logical connections between the scientific concepts guiding a hypothesis and the design of an experiment. They should demonstrate appropriate procedures, a knowledge base, and conceptual understanding of scientific investigations.

SAI 1f: Communicate and defend a scientific argument. Students in school science programs should develop the abilities associated with accurate and effective communication. These include writing and following procedures, expressing concepts, reviewing information, summarizing data, using language appropriately, developing diagrams and charts, explaining statistical analysis, speaking clearly and logically, constructing a reasoned argument, and responding appropriately to critical comments.

ST 1e: Communicate the problem, process, and solution. Students should present their results to students, teachers, and others in a variety of ways, such as orally, in writing, and in other forms--including models, diagrams, and demonstrations.

UCP 2: Evidence consists of observations and data on which to base scientific explanations. Using evidence to understand interactions allows individuals to predict changes in natural and designed systems. Models are tentative schemes or structures that correspond to real objects, events, or classes of events, and that have explanatory power. Models help scientists and engineers understand how things work. Models take many forms, including physical objects, plans, mental constructs, mathematical equations, and computer simulations.

UCP 3: Although most things are in the process of becoming different–changing–some properties of objects and processes are characterized by constancy, including the speed of light, the charge of an electron, and the total mass plus energy in the universe. Changes

KEY **SE** = Student Edition **TE** = Teacher Edition **CRF** = Chapter Resource File

might occur, for example, in properties of materials, position of objects, motion, and form and function of systems. Interactions within and among systems result in change. Changes vary in rate, scale, and pattern, including trends and cycles.

UCP 4: Evolution is a series of changes, some gradual and some sporadic, that accounts for the present form and function of objects, organisms, and natural and designed systems. The general idea of evolution is that the present arises from materials and forms of the past. Although evolution is most commonly associated with the biological theory explaining the process of descent with modification of organisms from common ancestors, evolution also describes changes in the universe.

FOCUS *(5 minutes)*

__ **Overview, TE** Review the objectives listed in the Student Edition. **(GENERAL)**

__ **PowerPoint® Resources** Use the customizable presentation to enhance your lecture.

__ **Bellringer, TE** Show students a picture of a coastline and have them draw pictures of what they think the shoreline will look like in 50 years and in 500 years. **(GENERAL)**

__ **Bellringer Transparency** Use this transparency as students enter the classroom and find their seats.

MOTIVATE *(10 minutes)*

__ **Group Activity, How Storms Threaten Shores, TE** In groups, students research how major storms threaten beaches and shorelines. **(GENERAL)**

TEACH *(65 minutes)*

__ **Directed Reading, CRF** These worksheets reinforce basic concepts and vocabulary presented in the section. **(BASIC)**

__ **Using the Figure, Breaking Waves, SE** Students use the photograph showing a wave breaking on a rock to determine where erosion is occurring in the photo. **(GENERAL)**

__ **Internet Activity, Landform Photographs, TE** Students research the Internet for examples of wave erosion. **(GENERAL)**

__ **Internet Activity, Landform Photographs, CRF** This worksheet directs student research for the topic presented in the Teacher Edition. **(GENERAL)**

__ **Reading Skill Builder, Reading Organizer, TE** Students are encouraged to take notes while reading this section. **(BASIC)**

__ **Teaching Transparency 90, Wave Erosion and Landforms with Worksheet** Use this Teaching Transparency and the accompanying worksheet to support your classroom discussion of wave erosion and landforms.

__ **Skill Builder, Writing, TE** Students write a story about how a grain of sand or pebble becomes part of a beach. **(BASIC)**

__ **Inquiry Lab, Beaches, SE** Students examine how wave action may change the shape of beaches and how these changes can be reduced. **(GENERAL)**

KEY **SE** = Student Edition **TE** = Teacher Edition **CRF** = Chapter Resource File

__ **Datasheet for Chapter Lab, Inquiry Lab, Beaches, CRF** Students use the datasheet to complete the Chapter Lab. **(GENERAL)**

__ **Interactive Tutor CD-ROM, Physical and Chemical Weathering** Students practice what they have learned using this interactive activity. **(GENERAL)**

CLOSE *(10 minutes)*

__ **Section Review, SE** Students answer end-of-section vocabulary, key ideas, critical thinking, and concept mapping questions. **(GENERAL)**

__ **Homework, Beach Composition, TE** Students research the composition of sand beaches around the world. **(ADVANCED)**

__ **Reteaching, Vocabulary Flash Cards, TE** Students create flashcards of the 12 landforms described in this section. **(BASIC)**

__ **Quiz, TE** Students answer 3 questions about this section. **(GENERAL)**

__ **Section Quiz, CRF** Use this quiz to assess students' understanding of the section. **(GENERAL) Also in Spanish**

__ **Spanish Section Quiz, Spanish Assessments** Use this quiz to assess students' understanding of the section. **(GENERAL)**

__ **Alternative Assessment, Landform Identification, TE** Students study photographs and identify the coastal formations shown. **(GENERAL)**

KEY **SE** = Student Edition **TE** = Teacher Edition **CRF** = Chapter Resource File

Lesson Plan

Section 3 Coastal Erosion and Deposition

PACING

Regular Schedule:	**with lab(s): N/A**	**without lab(s): 2 days**
Block Schedule:	**with lab(s): N/A**	**without lab(s): 1 day**

OBJECTIVES

1. **Explain** how changes in sea level affect coastlines.
2. **Describe** the features of a barrier island.
3. **Analyze** the effect of human activity on coastal land.

NATIONAL SCIENCE EDUCATION STANDARDS

ES 3c: Interactions among the solid earth, the oceans, the atmosphere, and organisms have resulted in the ongoing evolution of the earth system. We can observe some changes such as earthquakes and volcanic eruptions on a human time scale, but many processes such as mountain building and plate movements take place over hundreds of millions of years.

SPSP 5c: Some hazards, such as earthquakes, volcanic eruptions, and severe weather, are rapid and spectacular. But there are slow and progressive changes that also result in problems for individuals and societies. For example, change in stream channel position, erosion of bridge foundations, sedimentation in lakes and harbors, coastal erosions, and continuing erosion and wasting of soil and landscapes can all negatively affect society.

UCP 3: Although most things are in the process of becoming different—changing—some properties of objects and processes are characterized by constancy, including the speed of light, the charge of an electron, and the total mass plus energy in the universe. Changes might occur, for example, in properties of materials, position of objects, motion, and form and function of systems. Interactions within and among systems result in change. Changes vary in rate, scale, and pattern, including trends and cycles.

UCP 4: Evolution is a series of changes, some gradual and some sporadic, that accounts for the present form and function of objects, organisms, and natural and designed systems. The general idea of evolution is that the present arises from materials and forms of the past. Although evolution is most commonly associated with the biological theory explaining the process of descent with modification of organisms from common ancestors, evolution also describes changes in the universe.

FOCUS *(5 minutes)*

__ **Overview, TE** Review the objectives listed in the Student Edition. **(GENERAL)**

__ **PowerPoint® Resources** Use the customizable presentation to enhance your lecture.

__ **Bellringer, TE** Students are asked how life on Earth would be different if the sea level was 420 feet lower than it is today. **(GENERAL)**

KEY **SE** = Student Edition **TE** = Teacher Edition **CRF** = Chapter Resource File

Lesson Plan

__ **Bellringer Transparency** Use this transparency as students enter the classroom and find their seats.

MOTIVATE *(10 minutes)*

__ **Using the Figure, Sea Level Changes, TE** Use the figure showing sea level changes to lead a discussion. **(GENERAL)**

TEACH *(20 minutes)*

__ **Directed Reading, CRF** These worksheets reinforce basic concepts and vocabulary presented in the section. **(BASIC)**

__ **Quick Lab, Graphing Tides, SE** Students record daily tidal data for one month and graph the results. **(GENERAL)**

__ **Datasheet for Quick Lab, Graphing Tides, CRF** Students use the datasheet to complete the Quick Lab. **(GENERAL)**

__ **Teaching Transparency 91, Submergent Coastlines with Worksheet** Use this Teaching Transparency and the accompanying worksheet to support your classroom discussion of submergent coastlines.

__ **Activity, Poster Project, TE** Students research tides and create posters showing the relative positions of the sun, moon and Earth. **(ADVANCED)**

__ **Skill Builder, Writing, TE** Students create bumper stickers promoting coastal preservation. **(BASIC)**

__ **Geography Connection, Coastal Features, TE** Students research fiords, estuaries, and lagoons and create posters illustrating their findings. **(GENERAL)**

__ **Maps in Action, Coastal Erosion Near the Beaufort Sea,** Students enhance their map-reading skills by answering questions about a graphic map showing coastal erosion near the Beaufort Sea. **(GENERAL)**

__ **Teaching Transparency with Worksheet, Coastal Erosion Near the Beaufort Sea** Use this Teaching Transparency and the accompanying worksheet to help students perform the Maps in Action activity.

__ **Impact on Society, The Flooding of Venice, SE** This article discusses Venice's propensity to flood and the damages caused by erosion and rising sea levels.

__ **Group Activity, Debate Project, TE** Students research and debate MOSE, a project designed as a means of controlling flooding in Venice. **(BASIC)**

__ **Interactive Tutor CD-ROM, Coastal Zone** Students practice what they have learned using this interactive activity. **(GENERAL)**

__ **Making Models Lab, Erosion of a Submerging Coastal Profile, CRF** Students model a coastline to see how it is affected by increasing water levels. **(GENERAL)**

KEY **SE** = Student Edition **TE** = Teacher Edition **CRF** = Chapter Resource File

CLOSE *(10 minutes)*

__ **Section Review, SE** Students answer end-of-section vocabulary, key ideas, critical thinking, and concept mapping questions. **(GENERAL)**

__ **Homework, Coastal Landforms, TE** Students research characteristics of erosional and depositional coastal landforms. **(ADVANCED)**

__ **Reteaching, Sea-Level Changes, TE** Students create charts summarizing relative sea-level changes. **(BASIC)**

__ **Quiz, TE** Students answer 2 questions about this section. **(GENERAL)**

__ **Section Quiz, CRF** Use this quiz to assess students' understanding of the section. **(GENERAL) Also in Spanish**

__ **Spanish Section Quiz, Spanish Assessments** Use this quiz to assess students' understanding of the section. **(GENERAL)**

__ **Alternative Assessment, Landform Description, TE** Students read imaginary headlines and write a news story to accompany each headline. **(GENERAL)**

KEY SE = Student Edition TE = Teacher Edition CRF = Chapter Resource File

Lesson Plan

End of Chapter Review and Assessment

PACING

Regular Schedule: **with lab(s): N/A** **without lab(s): 2 days**
Block Schedule: **with lab(s): N/A** **without lab(s): 1 day**

__ **Chapter Review, SE** Students answer end-of-chapter vocabulary, key ideas, critical thinking, math, writing, and interpreting graphics questions. **(GENERAL)**

__ **Alternative Assessment, Quiz Show, TE** Help students study by creating a quiz-show style game. **(GENERAL)**

__ **Standardized Test Prep, SE** Have students complete the standardized test prep to help them prepare for standardized tests. **(GENERAL)**

__ **Holt PuzzlePro®** Use this software to create crossword puzzles and word searches that make learning vocabulary fun. **(GENERAL)**

__ **Concept Review, CRF** Use this worksheet to assess student's understanding of the main concepts presented in the chapter. **(GENERAL)**

__ **Critical Thinking, CRF** Students use critical thinking skills to answer questions about erosion by wind and waves. **(GENERAL)**

__ **Math Skills, CRF** This worksheet helps reinforce students' understanding of Order of Operations. **(GENERAL)**

__ **Graphing Skills, CRF** This worksheet helps reinforce students' understanding of Bar Graphs. **(GENERAL)**

__ **Study Guide, Concept Review, Erosion by Wind and Waves** Use this worksheet to assess student's understanding of the main concepts presented in the chapter. **(GENERAL)**

__ **Concept Review, Spanish Study Guide, Erosion by Wind and Waves** This worksheet helps reinforce English-language learners' understanding of the main concepts presented in the chapter. **(GENERAL)**

__ **Chapter Test A, CRF** Assign questions from the appropriate test for chapter assessment. **(GENERAL)**

__ **Chapter Test B, CRF** Assign questions from the appropriate test for chapter assessment. **(ADVANCED)**

__ **Spanish Chapter Test A, Spanish Assessments** Assign questions from the appropriate test for chapter assessment. **(GENERAL)**

__ **Test Generator, One-Stop Planner** Create a customized homework assignment, quiz, or test using the HRW Test Generator program. **(GENERAL)**

__ **Test Item Listing, CRF** Use the Test Item Listing to identify questions for a customized homework, quiz, or test.

__ **PowerPoint® Resources** Use the customizable presentation to help students practice standardized test-taking skills.

KEY **SE** = Student Edition **TE** = Teacher Edition **CRF** = Chapter Resource File

OTHER RESOURCE OPTIONS

__ **HRW Earth Science Video, Segment 9, "Weathering and Erosion"** Weathering and erosion, which continually reshape the landscape of Earth, are the focuses of this segment. **(GENERAL)**

__ **Brain Food Video Quiz** Use this video resource to assess student's progress. **(GENERAL)**

KEY **SE** = Student Edition **TE** = Teacher Edition **CRF** = Chapter Resource File

Lesson Plan

Chapter 19 The Ocean Basins

Chapter Opener

__ **Pre-Reading Activity, Booklet, SE**
__ **Using the Figure, The Ora Verde, TE (GENERAL)**

Section 1 The Water Planet

PACING

Regular Schedule:	**with lab(s): N/A**	**without lab(s): 2 days**
Block Schedule:	**with lab(s): N/A**	**without lab(s): 1 day**

OBJECTIVES

1. **Name** the major divisions of the global ocean.
2. **Describe** how oceanographers study the ocean.
3. **Explain** how sonar works.

NATIONAL SCIENCE EDUCATION STANDARDS

SAI 2c: Scientists rely on technology to enhance the gathering and manipulation of data. New techniques and tools provide new evidence to guide inquiry and new methods to gather data, thereby contributing to the advance of science. The accuracy and precision of the data, and therefore the quality of the exploration, depends on the technology used.

FOCUS *(5 minutes)*

__ **Overview, TE** Review the objectives listed in the Student Edition. **(GENERAL)**

__ **PowerPoint® Resources** Use the customizable presentation to enhance your lecture.

__ **Bellringer, TE** Students are asked why they think the science fiction author Arthur C. Clarke once remarked: "How inappropriate to call this planet Earth, when clearly it is Ocean.,, **(GENERAL)**

__ **Bellringer Transparency** Use this transparency as students enter the classroom and find their seats.

MOTIVATE *(10 minutes)*

__ **Using the Figure, Global Oceans, TE** Students study a map of the divisions of the global ocean to determine the areas of Earth covered by each ocean. **(GENERAL)**

TEACH *(20 minutes)*

__ **Directed Reading, CRF** These worksheets reinforce basic concepts and vocabulary presented in the section. **(BASIC)**

KEY **SE** = Student Edition **TE** = Teacher Edition **CRF** = Chapter Resource File

__ **Teaching Transparency 93, The Global Ocean with Worksheet** Use this Teaching Transparency and the accompanying worksheet to support your classroom discussion of the global ocean.

__ **Skill Builder, Writing, TE** Students research ocean expeditions and write articles detailing their findings. **(ADVANCED)**

__ **Reading Skill Builder, Paired Summarizing, SE** Students read the text in pairs and take turns summarizing what they have read. **(BASIC)**

__ **Language Arts Connection, 20,000 Leagues Under the Sea Today, TE** Students read excerpts from Jules Verne's novel and rewrite a scene. **(ADVANCED)**

__ **Quick Lab, Sonar, SE** Students learn about sonar by measuring and calculating compression waves. **(GENERAL)**

__ **Datasheet for Quick Lab, Sonar, CRF** Students use the datasheet to complete the Quick Lab. **(GENERAL)**

__ **Skill Builder, Math, TE** Students calculate the depth of the ocean floor. **(BASIC)**

__ **Biology Connection, Seeing with Sound, TE** Students review how animals use sonar or *echolocation*. **(GENERAL)**

__ **Career Focus, Oceanographer, SE** This article discusses some of the responsibilities and benefits of being an oceanographer. **(GENERAL)**

__ **Activity, Job Ad, TE** Students write a classified ad for a job aboard the research vessel *JOIDES Resolution*. **(GENERAL)**

CLOSE *(10 minutes)*

__ **Section Review, SE** Students answer end-of-section vocabulary, key ideas, critical thinking, and concept mapping questions. **(GENERAL)**

__ **Reteaching, Outline, TE** Students create a timeline of the advances in oceanographic research techniques. **(BASIC)**

__ **Quiz, TE** Students answer 3 questions about this section. **(GENERAL)**

__ **Section Quiz, CRF** Use this quiz to assess students' understanding of the section. **(GENERAL) Also in Spanish**

__ **Spanish Section Quiz, Spanish Assessments** Use this quiz to assess students' understanding of the section. **(GENERAL)**

__ **Alternative Assessment, Oceanographic Interview, TE** In pairs, students answer the question, "What is the most important advance in oceanographic research in the last 100 years?„ **(GENERAL)**

KEY **SE** = Student Edition **TE** = Teacher Edition **CRF** = Chapter Resource File

Lesson Plan

Section 2 Features of the Ocean Floor

PACING

Regular Schedule: **with lab(s): N/A** **without lab(s): 2 days**
Block Schedule: **with lab(s): N/A** **without lab(s): 1 day**

OBJECTIVES

1. **Describe** the main features of the continental margins.
2. **Describe** and the main features of the deep-ocean basin.

NATIONAL SCIENCE EDUCATION STANDARDS

[No standards correlate.]

FOCUS *(5 minutes)*

__ **Overview, TE** Review the objectives listed in the Student Edition. **(GENERAL)**

__ **PowerPoint® Resources** Use the customizable presentation to enhance your lecture.

__ **Bellringer, TE** Students make a sketch of what they think the ocean floor looks like from the Mid-Atlantic United States to Spain. **(GENERAL)**

__ **Bellringer Transparency** Use this transparency as students enter the classroom and find their seats.

MOTIVATE *(10 minutes)*

__ **Using the Figure, Modeling the Ocean Floor, TE** Students use the diagram to build models of the ocean floor. **(GENERAL)**

TEACH *(20 minutes)*

__ **Directed Reading, CRF** These worksheets reinforce basic concepts and vocabulary presented in the section. **(BASIC)**

__ **Graphic Organizer, Spider Map, SE** Use the graphic organizer to help students develop their analytical skills. **(GENERAL)**

__ **Teaching Transparency 94, Features of the Ocean Floor with Worksheet** Use this Teaching Transparency and the accompanying worksheet to support your classroom discussion of the features of the ocean floor.

__ **Making Models Lab, Island to Guyot, CRF** Students use modeling clay and water to model the transformation of a volcanic island into a guyot. **(GENERAL)**

KEY **SE** = Student Edition **TE** = Teacher Edition **CRF** = Chapter Resource File

Lesson Plan

CLOSE *(10 minutes)*

__ **Section Review, SE** Students answer end-of-section vocabulary, key ideas, critical thinking, and concept mapping questions. **(GENERAL)**

__ **Reteaching, Diagram, TE** Students label a diagram of the ocean floor with the features covered in the section. **(BASIC)**

__ **Quiz, TE** Students answer 4 questions about this section. **(GENERAL)**

__ **Section Quiz, CRF** Use this quiz to assess students' understanding of the section. **(GENERAL) Also in Spanish**

__ **Spanish Section Quiz, Spanish Assessments** Use this quiz to assess students' understanding of the section. **(GENERAL)**

__ **Alternative Assessment, Deep Sea Tour, TE** Students design travel brochures for a tour of the ocean floor. **(ADVANCED)**

KEY **SE** = Student Edition **TE** = Teacher Edition **CRF** = Chapter Resource File

Lesson Plan

Section 3 Ocean-Floor Sediments

PACING

Regular Schedule: **with lab(s): 2 days** **without lab(s): 1 day**
Block Schedule: **with lab(s): 1 day** **without lab(s): 0.5 day**

OBJECTIVES

1. **Describe** the formation of ocean-floor sediments.
2. **Explain** how ocean-floor sediments are classified by their physical composition.

NATIONAL SCIENCE EDUCATION STANDARDS

SAI 2c: Scientists rely on technology to enhance the gathering and manipulation of data. New techniques and tools provide new evidence to guide inquiry and new methods to gather data, thereby contributing to the advance of science. The accuracy and precision of the data, and therefore the quality of the exploration, depends on the technology used.

UCP 4: Evolution is a series of changes, some gradual and some sporadic, that accounts for the present form and function of objects, organisms, and natural and designed systems. The general idea of evolution is that the present arises from materials and forms of the past. Although evolution is most commonly associated with the biological theory explaining the process of descent with modification of organisms from common ancestors, evolution also describes changes in the universe.

FOCUS *(5 minutes)*

__ **Overview, TE** Review the objectives listed in the Student Edition. **(GENERAL)**

__ **PowerPoint® Resources** Use the customizable presentation to enhance your lecture.

__ **Bellringer, TE** Students are asked: "Where does sediment at the bottom of the ocean come from and what is it like?„ **(GENERAL)**

__ **Bellringer Transparency** Use this transparency as students enter the classroom and find their seats.

MOTIVATE *(10 minutes)*

__ **Demonstration, Sediments Sinking, TE** Students predict how various sediments will settle out if suspended in water. **(GENERAL)**

TEACH *(65 minutes)*

__ **Directed Reading, CRF** These worksheets reinforce basic concepts and vocabulary presented in the section. **(BASIC)**

__ **Discussion, Sediment Loading, TE** Students discuss how changes in land use could influence the amount of sediment loads that enter the ocean. **(GENERAL)**

KEY **SE** = Student Edition **TE** = Teacher Edition **CRF** = Chapter Resource File

__ **Demonstration, Turbidity Currents and Submarine Canyons, TE** Students learn how turbidity currents form submarine canyons. **(GENERAL)**

__ **Reading Skill Builder, Reading Organizer, TE** Students are encouraged to take notes while reading this section. **(BASIC)**

__ **Quick Lab, Diatoms, SE** Students observe diatoms under a microscope. **(GENERAL)**

__ **Datasheet for Quick Lab, Diatoms, CRF** Students use the datasheet to complete the Quick Lab. **(GENERAL)**

__ **Skills Practice Lab, Ocean-Floor Sediments, SE** Students use samples to determine the relationship between particle size and the settling rate of particles in water. **(GENERAL)**

__ **Datasheet for Chapter Lab, Skills Practice Lab, Ocean-Floor Sediments, CRF** Students use the datasheet to complete the Chapter Lab. **(GENERAL)**

__ **Maps in Action, Total Sediment Thickness of Earth's Oceans, SE** Students enhance their map-reading skills by answering questions about a graphic map showing the total sediment thickness of Earth's oceans. **(GENERAL)**

__ **Teaching Transparency 95, Total Sediment Thickness of Earth's Oceans with Worksheet** Use this Teaching Transparency and the accompanying worksheet to help students perform the Maps in Action activity.

__ **Discussion, Sediments of the Western Shore, TE** Students discuss the differences between sediments found on the eastern and western shores of North and South America. **(GENERAL)**

__ **Internet Activity, Deep Sea Cores, TE** Students use the internet to learn what scientists hope to find by studying deep sea sediment core samples. **(ADVANCED)**

__ **Internet Activity, Deep Sea Cores, CRF** This worksheet directs student research for the topic presented in the Teacher Edition. **(ADVANCED)**

__ **Activity, Job Ad, TE** Students write an ad for an ocean-related career. **(GENERAL)**

__ **Inquiry Lab, Sediment Loading, CRF** Students examine how changes in land use alter sediment loads in rivers and identify ways to decrease the impact on oceans. **(GENERAL)**

CLOSE *(10 minutes)*

__ **Section Review, SE** Students answer end-of-section vocabulary, key ideas, critical thinking, and concept mapping questions. **(GENERAL)**

__ **Reteaching, Peer Reviewing, TE** In pairs, students write questions about the material in this section and quiz each other. **(BASIC)**

__ **Quiz, TE** Students answer 3 questions about this section. **(GENERAL)**

__ **Section Quiz, CRF** Use this quiz to assess students' understanding of the section. **(GENERAL) Also in Spanish**

__ **Spanish Section Quiz, Spanish Assessments** Use this quiz to assess students' understanding of the section. **(GENERAL)**

__ **Alternative Assessment, Comparing and Contrasting Sediments, TE** Students compare and contrast the three types of sediment found on the sea floor. **(BASIC)**

KEY SE = Student Edition TE = Teacher Edition CRF = Chapter Resource File

End of Chapter Review and Assessment

PACING

Regular Schedule: **with lab(s): N/A** **without lab(s): 2 days**
Block Schedule: **with lab(s): N/A** **without lab(s): 1 day**

__ **Chapter Review, SE** Students answer end-of-chapter vocabulary, key ideas, critical; thinking, math, writing, and interpreting graphics questions. **(GENERAL)**

__ **Alternative Assessment, Debate, TE** In groups, students debate whether or not we know more about the moon than we do about Earth's oceans **(ADVANCED)**

__ **Standardized Test Prep, SE** Have students complete the standardized test prep to help them prepare for standardized tests. **(GENERAL)**

__ **Holt PuzzlePro®** Use this software to create crossword puzzles and word searches that make learning vocabulary fun. **(GENERAL)**

__ **Concept Review, CRF** Use this worksheet to assess student's understanding of the main concepts presented in the chapter. **(GENERAL)**

__ **Critical Thinking, CRF** Students use critical thinking skills to answer questions about ocean basins. **(GENERAL)**

__ **Math Skills, CRF** This worksheet helps reinforce students' understanding of Algebraic Rearrangements. **(GENERAL)**

__ **Graphing Skills, CRF** This worksheet helps reinforce students' understanding of Pie Graphs. **(GENERAL)**

__ **Study Guide, Concept Review, The Ocean Basins** Use this worksheet to assess student's understanding of the main concepts presented in the chapter. **(GENERAL)**

__ **Concept Review, Spanish Study Guide, The Ocean Basins** This worksheet helps reinforce English-language learners' understanding of the main concepts presented in the chapter. **(GENERAL)**

__ **Chapter Test A, CRF** Assign questions from the appropriate test for chapter assessment. **(GENERAL)**

__ **Chapter Test B, CRF** Assign questions from the appropriate test for chapter assessment. **(ADVANCED)**

__ **Spanish Chapter Test A, Spanish Assessments** Assign questions from the appropriate test for chapter assessment. **(GENERAL)**

__ **Test Generator, One-Stop Planner** Create a customized homework assignment, quiz, or test using the HRW Test Generator program. **(GENERAL)**

__ **Test Item Listing, CRF** Use the Test Item Listing to identify questions for a customized homework, quiz, or test.

__ **PowerPoint® Resources** Use the customizable presentation to help students practice standardized test-taking skills.

KEY **SE** = Student Edition **TE** = Teacher Edition **CRF** = Chapter Resource File

OTHER RESOURCE OPTIONS

__ **HRW Earth Science Video, Segment 14, "Features of the Ocean Floor"** The oceans comprise three-quarters of Earth's surface and contain some of the planet's largest topographic structures, such as Hawaii, the topic of this segment.

__ **Brain Food Video Quiz** Use this video resource to assess student's progress. **(GENERAL)**

KEY **SE** = Student Edition **TE** = Teacher Edition **CRF** = Chapter Resource File

Chapter 20 Ocean Water

__ **Pre-Reading Activity, Four-Corner Fold, SE**

__ **Using the Figure, Water of Life, TE**

__ **Long-Term Project, Water Clarity, Long-Term Projects** Students discover how changes in water clarity relate to changes in the environment. **(ADVANCED)**

Section 1 Properties of Ocean Water

PACING

Regular Schedule:	**with lab(s): 2 days**	**without lab(s): 1 day**
Block Schedule:	**with lab(s): 1 day**	**without lab(s): 0.5 day**

OBJECTIVES

1. **Describe** the chemical composition of ocean water.
2. **Describe** the salinity, temperature, density, and color of ocean water.

NATIONAL SCIENCE EDUCATION STANDARDS

SPSP 4a: Natural ecosystems provide an array of basic processes that affect humans. Those processes include maintenance of the quality of the atmosphere, generation of soils, control of the hydrologic cycle, disposal of wastes, and recycling of nutrients. Humans are changing many of these basic processes, and the changes may be detrimental to humans.

FOCUS *(5 minutes)*

__ **Overview, TE** Review the objectives listed in the Student Edition. **(GENERAL)**

__ **PowerPoint® Resources** Use the customizable presentation to enhance your lecture.

__ **Bellringer, TE** Students are asked to list everything they think is found in ocean water. **(GENERAL)**

__ **Bellringer Transparency** Use this transparency as students enter the classroom and find their seats.

MOTIVATE *(10 minutes)*

__ **Activity, Losing Fizz, TE** Students observe two samples of club soda and record their observations. **(GENERAL)**

KEY SE = Student Edition **TE** = Teacher Edition **CRF** = Chapter Resource File

TEACH *(65 minutes)*

__ **Directed Reading, CRF** These worksheets reinforce basic concepts and vocabulary presented in the section. **(BASIC)**

__ **Teaching Transparency 96, Dissolved Gases in the Ocean with Worksheet** Use this Teaching Transparency and the accompanying worksheet to support your classroom discussion of the dissolved gases in the ocean.

__ **Meteorology Connection, Positive Feedback, TE** Students draw diagrams or concept maps illustrating how greenhouse gases affect the atmosphere. **(ADVANCED)**

__ **Skill Builder, Math, TE** Students determine the mass of each element in a sample of ocean salts. **(GENERAL)**

__ **Quick Lab, Dissolving Solids, SE** Students observe how water temperature affects salinity. **(GENERAL)**

__ **Datasheet for Quick Lab, Dissolving Solids, CRF** Students use the datasheet to complete the Quick Lab. **(GENERAL)**

__ **Using the Figure, Salinity Around the World, TE** Students use the figure to discuss ocean salinity. **(GENERAL)**

__ **Teaching Transparency 97, Average Surface Salinity of the Global Ocean with Worksheet** Use this Teaching Transparency and the accompanying worksheet to support your classroom discussion of the average surface salinity of the global ocean.

__ **Activity, Ocean Surface Temperatures, TE** In groups, students record and compare the temperatures of water warmed by heat lamps. **(GENERAL)**

__ **Reading Skill Builder, Paired Summarizing, TE** In pairs, students read and summarize the description of the thermocline. **(BASIC)**

__ **Using the Figure, Thermoclines, TE** Students use the graph of a thermocline to review the axes of graphs. **(GENERAL)**

__ **Environmental Science Connection, Surface Water Temperatures, TE** Students research surface ocean currents along the coasts of the U.S. **(GENERAL)**

__ **Quick Lab, Density Factors, SE** Students examine how density differs between salt water and fresh water and cool water and warm water. **(GENERAL)**

__ **Datasheet for Quick Lab, Density Factors, CRF** Students use the datasheet to complete the Quick Lab. **(GENERAL)**

__ **Skills Practice Lab, Ocean Water Density, SE** Students observe the effects of temperature and salinity on the density of salt water. **(GENERAL)**

__ **Datasheet for Chapter Lab, Skills Practice Lab, Ocean Water Density, CRF** Students use the datasheet to complete the Chapter Lab. **(GENERAL)**

__ **Maps in Action, Sea Surface Temperatures in August, SE** Students enhance their map-reading skills by answering questions about a map showing global sea surface temperatures in the month of August. **(GENERAL)**

KEY **SE** = Student Edition **TE** = Teacher Edition **CRF** = Chapter Resource File

__ **Teaching Transparency 100, Sea Surface Temperatures in August with Worksheet** Use this Teaching Transparency and the accompanying worksheet to help students perform the Maps in Action activity.

__ **Using the Figure, Liquid Density,** TE Students use the figure illustrating liquids and densities to predict if a layer of salt water would move above or below a layer of fresh water. **(BASIC)**

__ **Skills Practice Lab, The Blue-Green Ocean, CRF** Students observe how varying amounts of sunlight affect phytoplankton growth. **(GENERAL)**

CLOSE *(10 minutes)*

__ **Section Review, SE** Students answer end-of-section vocabulary, key ideas, critical thinking, and concept mapping questions. **(GENERAL)**

__ **Homework, Freezing Fresh and Salt Water, TE** Students identify whether salt or fresh water freezes faster. **(BASIC)**

__ **Reteaching, Basic Diagrams, TE** Students create diagrams illustrating substances found in the oceans and their sources. **(BASIC)**

__ **Quiz, TE** Students answer 3 questions about this sectoin. **(GENERAL)**

__ **Section Quiz, CRF** Use this quiz to assess students' understanding of the section. **(GENERAL) Also in Spanish**

__ **Spanish Section Quiz, Spanish Assessments** Use this quiz to assess students' understanding of the section. **(GENERAL)**

__ **Alternative Assessment, Essay, TE** Students write essays connecting dissolved carbon dioxide in the oceans to atmospheric carbon dioxide, ocean temperature, and climate change. **(ADVANCED)**

KEY **SE** = Student Edition **TE** = Teacher Edition **CRF** = Chapter Resource File

Lesson Plan

Section 2 Life in the Oceans

PACING

Regular Schedule: **with lab(s): N/A** **without lab(s): 2 days**
Block Schedule: **with lab(s): N/A** **without lab(s): 1 day**

OBJECTIVES

1. **Explain** how marine organisms alter the chemistry of ocean water.
2. **Explain why** plankton can be called the *foundation of life in the ocean.*
3. **Describe** the major zones of life in the ocean.

NATIONAL SCIENCE EDUCATION STANDARDS

LS 4c: Organisms both cooperate and compete in ecosystems. The interrelationships and interdependencies of these organisms may generate ecosystems that are stable for hundreds or thousands of years.

FOCUS *(5 minutes)*

__ **Overview, TE** Review the objectives listed in the Student Edition. **(GENERAL)**

__ **PowerPoint® Resources** Use the customizable presentation to enhance your lecture.

__ **Bellringer, TE** Students are asked to make a list of marine organisms they are familiar with and identify whether the organisms live on the ocean floor or swim freely. **(GENERAL)**

__ **Bellringer Transparency** Use this transparency as students enter the classroom and find their seats.

MOTIVATE *(10 minutes)*

__ **Group Activity, Poster Project, TE** In groups, students create posters illustrating how marine organisms help maintain the chemical balance of ocean water. **(GENERAL)**

TEACH *(20 minutes)*

__ **Directed Reading, CRF** These worksheets reinforce basic concepts and vocabulary presented in the section. **(BASIC)**

__ **Reading Skill Builder, Anticipation Guide, TE** Students predict the relationship between ocean chemistry and marine life. **(BASIC)**

__ **Using the Figure, Upwelling, TE** Students use the figure illustrating prevailing surface winds to discuss what would happen if the prevailing winds shifted to blow perpendicular to the coastline. **(ADVANCED)**

KEY **SE** = Student Edition **TE** = Teacher Edition **CRF** = Chapter Resource File

__ **Teaching Transparency 98, Upwelling with Worksheet** Use this Teaching Transparency and the accompanying worksheet to support your classroom discussion of upwelling.

__ **Activity, Marine Food Webs, TE** Students research marine organisms that live in one ecosystem. **(GENERAL)**

__ **Graphic Organizer, Comparison Table, SE** Use the Graphic Organizer to help students develop their analytical skills. **(GENERAL)**

CLOSE *(10 minutes)*

__ **Section Review, SE** Students answer end-of-section vocabulary, key ideas, critical thinking, and concept mapping questions. **(GENERAL)**

__ **Reteaching, Peer Reviewing, TE** In pairs, students write five questions using material from this section. **(BASIC)**

__ **Quiz, TE** Students answer 2 questions about this section. **(GENERAL)**

__ **Section Quiz, CRF** Use this quiz to assess students' understanding of the section. **(GENERAL) Also in Spanish**

__ **Spanish Section Quiz, Spanish Assessments** Use this quiz to assess students' understanding of the section. **(GENERAL)**

__ **Alternative Assessment, Modeling the Ocean, TE** In groups, students create 3-dimensional models of the benthic and pelagic zones. **(GENERAL)**

KEY **SE** = Student Edition **TE** = Teacher Edition **CRF** = Chapter Resource File

Lesson Plan

Section 3 Ocean Resources

PACING

Regular Schedule: **with lab(s): N/A** **without lab(s): 2 days**
Block Schedule: **with lab(s): N/A** **without lab(s): 1 day**

OBJECTIVES

1. **Describe** three important resources of the ocean.
2. **Explain** the threat water pollution poses to marine organisms.

NATIONAL SCIENCE EDUCATION STANDARDS

LS 4e: Human beings live within the world's ecosystems. Increasingly, humans modify ecosystems as a result of population growth, technology, and consumption. Human destruction of habitats through direct harvesting, pollution, atmospheric changes, and other factors is threatening current global stability, and if not addressed, ecosystems will be irreversibly affected.

SPSP 3a: Human populations use resources in the environment in order to maintain and improve their existence. Natural resources have been and will continue to be used to maintain human populations.

FOCUS *(5 minutes)*

__ **Overview, TE** Review the objectives listed in the Student Edition. **(GENERAL)**

__ **PowerPoint® Resources** Use the customizable presentation to enhance your lecture.

__ **Bellringer, TE** Students are asked to list everything they think is found in ocean water. **(GENERAL)**

__ **Bellringer Transparency** Use this transparency as students enter the classroom and find their seats.

MOTIVATE *(10 minutes)*

__ **Discussion, Fresh Water from Salt Water, TE** In groups, students research desalination and outline key points for discussion.

TEACH *(20 minutes)*

__ **Directed Reading, CRF** These worksheets reinforce basic concepts and vocabulary presented in the section. **(BASIC)**

__ **Activity, Where's the Oil? TE** Students use the Web and other sources to learn where the world's oil reserves are and how much oil they produce. **(BASIC)**

__ **Skill Builder, Vocabulary, TE** Students learn the root of the word *petrol*. **(GENERAL)**

KEY **SE** = Student Edition **TE** = Teacher Edition **CRF** = Chapter Resource File

__ **Internet Activity, The Romance of Aquaculture, TE** Students research issues relevant to aquaculture and report to the class or carry out a project. **(ADVANCED)**

__ **Internet Activity, The Romance of Aquaculture, CRF** This worksheet directs student research for the topic presented in the Teacher Edition. **(GENERAL)**

__ **Eye on the Environment, A Harbor Makes a Comeback, SE** Boston Harbor's tragic history and clean-up efforts are discussed.

__ **Group Activity, Curing Pollution in Boston Harbor, TE** Students research the history of the pollution of Boston Harbor and present their findings to the class. **(ADVANCED)**

__ **Inquiry Lab, Oil Spill! CRF** Students use a pie pan, oil, and water to model an oil spill and design effective ways to contain and clean up oil spills. **(GENERAL)**

CLOSE *(10 minutes)*

__ **Section Review, SE** Students answer end-of-section vocabulary, key ideas, critical thinking, and concept mapping questions. **(GENERAL)**

__ **Reteaching, Basic Diagrams, TE** Have students create diagrams that illustrate the substances found in the oceans and their sources. **(BASIC)**

__ **Quiz, TE** Students answer 2 questions about this section. **(GENERAL)**

__ **Section Quiz, CRF** Use this quiz to assess students' understanding of the section. **(GENERAL) Also in Spanish**

__ **Spanish Section Quiz, Spanish Assessments** Use this quiz to assess students' understanding of the section. **(GENERAL)**

__ **Alternative Assessment, Debate, TE** Students research and engage in a debate on ocean oil drilling. **(ADVANCED)**

KEY **SE** = Student Edition **TE** = Teacher Edition **CRF** = Chapter Resource File

Lesson Plan

End of Chapter Review and Assessment

PACING

Regular Schedule:	**with lab(s): N/A**	**without lab(s): 2 days**
Block Schedule:	**with lab(s): N/A**	**without lab(s): 1 day**

__ **Chapter Review, SE** Students answer end-of-chapter vocabulary, key ideas, critical thinking, math, writing, and interpreting graphics questions. **(GENERAL)**

__ **Alternative Assessment, Ocean Video, TE** In groups, students create plans for an animated video or film about the concepts covered in this chapter. **(ADVANCED)**

__ **Standardized Test Prep, SE** Have students complete the standardized test prep to help them prepare for standardized tests. **(GENERAL)**

__ **Holt PuzzlePro®** Use this software to create crossword puzzles and word searches that make learning vocabulary fun. **(GENERAL)**

__ **Concept Review, CRF** Use this worksheet to assess student's understanding of the main concepts presented in the chapter. **(GENERAL)**

__ **Critical Thinking, CRF** Students use critical thinking skills to answer questions about ocean water. **(GENERAL)**

__ **Math Skills, CRF** This worksheet helps reinforce students' understanding of Scientific Notation. **(GENERAL)**

__ **Graphing Skills, CRF** This worksheet helps reinforce students' understanding of Pie graphs. **(GENERAL)**

__ **Concept Review, Study Guide, Plate Tectonics** Use this worksheet to assess student's understanding of the main concepts presented in the chapter. **(GENERAL)**

__ **Concept Review, Spanish Study Guide, Plate Tectonics** This worksheet helps reinforce English-language learners' understanding of the main concepts presented in the chapter. **(GENERAL)**

__ **Chapter Test A, CRF** Assign questions from the appropriate test for chapter assessment. **(GENERAL)**

__ **Chapter Test B, CRF** Assign questions from the appropriate test for chapter assessment. **(ADVANCED)**

__ **Spanish Chapter Test A, Spanish Assessments** Assign questions from the appropriate test for chapter assessment. **(GENERAL)**

__ **Test Generator, One-Stop Planner** Create a customized homework assignment, quiz, or test using the HRW Test Generator program. **(GENERAL)**

__ **Test Item Listing, CRF** Use the Test Item Listing to identify questions for a customized homework, quiz, or test.

__ **PowerPoint® Resources** Use the customizable presentation to help students practice standardized test-taking skills.

KEY SE = Student Edition **TE** = Teacher Edition **CRF** = Chapter Resource File

OTHER RESOURCE OPTIONS

_ **NOVA Video, Segment 1, "Adrift on the Gulf Stream"** NOVA explores the Gulf Stream's importance to ocean life and climate from a variety of vantage points.

__ **Brain Food Video Quiz** Use this video resource to assess student's progress. **(GENERAL)**

KEY **SE** = Student Edition **TE** = Teacher Edition **CRF** = Chapter Resource File

Chapter 21 Movements of the Ocean

Chapter Opener

__ **Pre-Reading Activity, Layered Book, SE**

__ **Using the Figure, Sunset Tide, TE (GENERAL)**

__ **Long-Term Project, Tides at the Shoreline, Long-Term Projects** Students observe the changes and patterns of the tidal range of their area over a period of two months. **(ADVANCED)**

Section 1 Ocean Currents

PACING

Regular Schedule:	**with lab(s): N/A**	**without lab(s): 2 days**
Block Schedule:	**with lab(s): N/A**	**without lab(s): 1 day**

OBJECTIVES

1. **Describe** how wind patterns, the rotation of Earth, and continental barriers affect surface currents in the ocean.
2. **Identify** the major factor that determines the direction in which a surface current circulates.
3. **Explain** how differences in the density of ocean water affect the flow of deep currents.

NATIONAL SCIENCE EDUCATION STANDARDS

ES 1c: Heating of earth's surface and atmosphere by the sun drives convection within the atmosphere and oceans, producing winds and ocean currents

FOCUS *(5 minutes)*

__ **Overview, TE** Review the objectives listed in the Student Edition. **(GENERAL)**

__ **PowerPoint® Resources** Use the customizable presentation to enhance your lecture.

__ **Bellringer, TE** Students are asked to think about movies or pictures they have seen that shown an ocean or a beach. **(GENERAL)**

__ **Bellringer Transparency** Use this transparency as students enter the classroom and find their seats.

KEY **SE** = Student Edition **TE** = Teacher Edition **CRF** = Chapter Resource File

MOTIVATE *(10 minutes)*

__ **Discussion, Beachcombing, TE** Students list things found on beaches and discuss where the objects may have come from. **(GENERAL)**

TEACH *(20 minutes)*

__ **Directed Reading, CRF** These worksheets reinforce basic concepts and vocabulary presented in the section. **(BASIC)**

__ **Demonstration, Modeling Ocean Currents, TE** Use an overhead projector, a clear container, water, food coloring, and a straw to simulate an ocean current. **(GENERAL)**

__ **Using the Figure, The Ocean Conveyor Belt, TE** Students are asked how the temperature of water currents is indicated on the world map. **(GENERAL)**

__ **Teaching Transparency 101, Major Surface Currents of Earth's Oceans with Worksheet** Use this Teaching Transparency and the accompanying worksheet to support your classroom discussion of major surface currents of Earth's oceans.

__ **Reading Skill Builder, Paired Summarizing, TE** Students read the text in pairs and take turns summarizing what they have read. **(BASIC)**

__ **Geography Connection, Research, TE** Students compare the weather and precipitation of two locations in the same latitude. **(ADVANCED)**

__ **Quick Lab, Ocean Currents, SE** Students learn about ocean currents by using a plate, water, paper confetti, and a straw. **(GENERAL)**

__ **Datasheet for Quick Lab, Ocean Currents, CRF** Students use the datasheet to complete the Quick Lab. **(GENERAL)**

__ **Skill Builder, Vocabulary, TE** Students review the roots of the words *salinity*, *density*, and *turbidity*. **(GENERAL)**

__ **History Connection, Research, TE** Students research Benjamin Franklin's maritime observations. **(GENERAL)**

__ **Graphic Organizer, Comparison Table, SE** Use the graphic organizer to help students develop their analytical skills. **(GENERAL)**

__ **Biology Connection, Vertical Ocean Movements, TE** Students review upward and downward motions of ocean water. **(GENERAL)**

__ **Maps in Action, Roaming Rubber Duckies, SE** Students enhance their map-reading skills by answering questions about a graphic map showing the estimated route taken by bathtub toys spilled from a cargo ship. **(GENERAL)**

__ **Teaching Transparency 105, Roaming Rubber Duckies with Worksheet** Use this Teaching Transparency and the accompanying worksheet to help students perform the Maps in Action activity.

__ **Group Activity, Poster Project, TE** Students create posters showing how they would find objects lost in a cargo spill. **(GENERAL)**

__ **Internet Activity, Beachcomber Tales, TE** Students research how floating objects have helped to understand ocean currents. **(ADVANCED)**

KEY **SE** = Student Edition **TE** = Teacher Edition **CRF** = Chapter Resource File

__ **Internet Activity, Beachcomber Tales, CRF** This worksheet directs student research for the topic presented in the Teacher Edition. **(ADVANCED)**

__ **Science and Technology, Energy from the Ocean, SE** Methods for using ocean features to generate electricity are discussed.

__ **Activity, Exploring Water Power, TE** Students build models of a water wheel. **(ADVANCED)**

__ **Group Activity, Power Plants, TE** Students draw diagrams showing how the mechanical motion of turbines generates electricity. **(ADVANCED)**

__ **Skills Practice Lab, Ocean Currents and Water Temperature, CRF** Students use food coloring, ice, and water to see how ocean currents move in layers. **(GENERAL)**

__ **Interactive Tutor CD-ROM, Heating the Ocean** Students practice what they have learned using this interactive activity. **(GENERAL)**

__ **Interactive Tutor CD-ROM, Ocean Circulation** Students practice what they have learned using this interactive activity. **(GENERAL)**

CLOSE *(10 minutes)*

__ **Section Review, SE** Students answer end-of-section vocabulary, key ideas, critical thinking, and concept mapping questions. **(GENERAL)**

__ **Reteaching, TE** Students create a Venn diagram to compare surface currents and deep ocean currents. **(BASIC)**

__ **Quiz, TE** Students answer 2 questions about this section. **(GENERAL)**

__ **Section Quiz, CRF** Use this quiz to assess students' understanding of the section. **(GENERAL) Also in Spanish**

__ **Spanish Section Quiz, Spanish Assessments** Use this quiz to assess students' understanding of the section. **(GENERAL)**

__ **Alternative Assessment, Adventures of a Drop, TE** Students write stories about a drop of water traveling through the ocean. **(ADVANCED)**

KEY **SE** = Student Edition **TE** = Teacher Edition **CRF** = Chapter Resource File

Lesson Plan

Section 2 Ocean Waves

PACING

Regular Schedule: **with lab(s): 2 days** **without lab(s): 1 day**
Block Schedule: **with lab(s): 1 day** **without lab(s): 0.5 day**

OBJECTIVES

1. **Describe** the formation of waves and the factors that affect wave size.
2. **Explain** how waves interact with the coastline.
3. **Identify** the cause of destructive ocean waves.

NATIONAL SCIENCE EDUCATION STANDARDS

PS 6a: Waves, including sound and seismic waves, waves on water, and light waves, have energy and can transfer energy when they interact with matter.

FOCUS *(5 minutes)*

__ **Overview, TE** Review the objectives listed in the Student Edition. **(GENERAL)**

__ **PowerPoint® Resources** Use the customizable presentation to enhance your lecture.

__ **Bellringer, TE** Have students list as many words as they can think of that describe the motion of ocean waves. **(GENERAL)**

__ **Bellringer Transparency** Use this transparency as students enter the classroom and find their seats.

MOTIVATE *(10 minutes)*

__ **Demonstration, Student Wave, TE** Students stand side-by-side to demonstrate the motion of an ocean wave. **(GENERAL)**

TEACH *(65 minutes)*

__ **Directed Reading, CRF** These worksheets reinforce basic concepts and vocabulary presented in the section. **(BASIC)**

__ **Using the Figure, Comprehension Check, TE** Students use the figure of a bottle and a wave to review why objects bob up and down in waves. **(BASIC)**

__ **Group Activity, Making Waves, TE** Students use coiled metal springs to create waves. **(GENERAL)**

__ **Teaching Transparency 102, Wave Motion and Wave Energy with Worksheet** Use this Teaching Transparency and the accompanying worksheet to support your classroom discussion of wave motion and wave energy.

__ **Skill Builder, Math, TE** Students calculate the speed of a wave. **(GENERAL)**

KEY **SE** = Student Edition **TE** = Teacher Edition **CRF** = Chapter Resource File

__ **Quick Lab, Waves, SE** Students learn about waves by using a rectangular pan, a cork, tape, and a spoon. **(GENERAL)**

__ **Datasheet for Quick Lab, Waves, CRF** Students use the datasheet to complete the Quick Lab. **(GENERAL)**

__ **Reading Skill Builder, Anticipation Guide, TE** Students predict what factors determine the size of a wave before reading the page. **(BASIC)**

__ **Activity, Wave Refraction in a Tank, TE** In groups, students explore the effects of waves in shallow water. **(GENERAL)**

__ **Teaching Transparency 103, The Formation of Breakers with Worksheet** Use this Teaching Transparency and the accompanying worksheet to support your classroom discussion of the formation of breakers.

__ **Discussion, Water Safety, TE** Students explain what a rip current is, how to identify one, and how to protect themselves from the dangerous current. **(BASIC)**

__ **Making Models Lab, Wave Motion, SE** Students simulate wave motion and observe how energy generates wave motion in water. **(GENERAL)**

__ **Datasheet for Chapter Lab, Making Models Lab, Wave Motion, CRF** Students use the datasheet to complete the Chapter Lab. **(GENERAL)**

__ **Inquiry Lab, Tsunami! CRF** Students model tsunami wave movement to see which types of shorelines are most affected by tsunamis. **(GENERAL)**

CLOSE *(10 minutes)*

__ **Section Review, SE** Students answer end-of-section vocabulary, key ideas, critical thinking, and concept mapping questions. **(GENERAL)**

__ **Homework, Longshore Currents, TE** Students research and write reports on longshore currents and their effects on the environment. **(ADVANCED)**

__ **Reteaching, Learn by Teaching, TE** In groups, students prepare presentations explaining how wind forms ocean waves and how waves affect coastlines and beaches. **(BASIC)**

__ **Quiz, TE** Students answer 4 questions about this section. **(GENERAL)**

__ **Section Quiz, CRF** Use this quiz to assess students' understanding of the section. **(GENERAL) Also in Spanish**

__ **Spanish Section Quiz, Spanish Assessments** Use this quiz to assess students' understanding of the section. **(GENERAL)**

__ **Alternative Assessment, Wave Haikus, TE** Students write a haiku about waves. **(GENERAL)**

KEY **SE** = Student Edition **TE** = Teacher Edition **CRF** = Chapter Resource File

Section 3 Tides

PACING

Regular Schedule:	**with lab(s): N/A**	**without lab(s): 2 days**
Block Schedule:	**with lab(s): N/A**	**without lab(s): 1 day**

OBJECTIVES

1. **Describe** how the gravitational pull of the moon causes tides.
2. **Compare** spring tides and neap tides.
3. **Describe** how tidal oscillations affect tidal patterns.
4. **Explain** how the coastline affects tidal currents.

NATIONAL SCIENCE EDUCATION STANDARDS

[None]

FOCUS *(5 minutes)*

__ **Overview, TE** Review the objectives listed in the Student Edition. **(GENERAL)**

__ **PowerPoint® Resources** Use the customizable presentation to enhance your lecture.

__ **Bellringer, TE** Have students answer the following questions: "What does the phrase 'high and dry' mean?" and "Where do you think this phrase comes from?" **(GENERAL)**

__ **Bellringer Transparency** Use this transparency as students enter the classroom and find their seats.

MOTIVATE *(10 minutes)*

__ **Using the Figure, Tidal Bulges, TE** Students use the figure of the moon and Earth to review tidal bulges. **(GENERAL)**

TEACH *(20 minutes)*

__ **Directed Reading, CRF** These worksheets reinforce basic concepts and vocabulary presented in the section. **(BASIC)**

__ **Using the Figure, Lunar Motions, TE** Students use the figure of spring and neap tides to identify the phases of the moon that would coincide with the highest tides. **(GENERAL)**

__ **Teaching Transparency 104, Spring Tides and Neap Tides with Worksheet** Use this Teaching Transparency and the accompanying worksheet to support your classroom discussion of spring tides and neap tides.

__ **Skill Builder, Graphing, TE** Students research high and low tides for a month and create a graph. **(ADVANCED)**

__ **Discussion, Bay of Fundy, TE** Students discuss why tides in Canada's Bay of Fundy are among the highest in the world. **(GENERAL)**

KEY **SE** = Student Edition **TE** = Teacher Edition **CRF** = Chapter Resource File

__ **Astronomy Connection, Same Face, TE** Using a compass, students demonstrate why the moon always shows the same side to people on Earth. **(GENERAL)**

__ **Interactive Tutor CD-ROM, Waves and Tides** Students practice what they have learned using this interactive activity. **(GENERAL)**

CLOSE *(10 minutes)*

__ **Section Review, SE** Students answer end-of-section vocabulary, key ideas, critical thinking, and concept mapping questions. **(GENERAL)**

__ **Reteaching, Tidal Games, TE** In pairs, students write questions and answers about causes of tides and tidal patterns. **(BASIC)**

__ **Quiz, TE** Students answer 4 questions about this section. **(GENERAL)**

__ **Section Quiz, CRF** Use this quiz to assess students' understanding of the section. **(GENERAL) Also in Spanish**

__ **Spanish Section Quiz, Spanish Assessments** Use this quiz to assess students' understanding of the section. **(GENERAL)**

__ **Alternative Assessment, Making Mobiles, TE** Students make mobiles showing the sun, Earth, and moon and how these three bodies cause tides to form. **(ADVANCED)**

KEY **SE** = Student Edition **TE** = Teacher Edition **CRF** = Chapter Resource File

End of Chapter Review and Assessment

PACING

Regular Schedule:	**with lab(s): N/A**	**without lab(s): 2 days**
Block Schedule:	**with lab(s): N/A**	**without lab(s): 1 day**

__ **Chapter Review, SE** Students answer end-of-chapter vocabulary, key ideas, critical thinking, math, writing, and interpreting graphics questions. **(GENERAL)**

__ **Alternative Assessment, [TITLE], TE** Students create models representing the movement of waves, tides, and currents and their effects on coastal areas. **(GENERAL)**

__ **Standardized Test Prep, SE** Have students complete the standardized test prep to help them prepare for standardized tests. **(GENERAL)**

__ **Holt PuzzlePro®** Use this software to create crossword puzzles and word searches that make learning vocabulary fun. **(GENERAL)**

__ **Concept Review, CRF** Use this worksheet to assess student's understanding of the main concepts presented in the chapter. **(GENERAL)**

__ **Critical Thinking, CRF** Students use this worksheet to develop and practice critical thinking skills. **(ADVANCED)**

__ **Math Skills, CRF** This worksheet helps reinforce students' understanding of significant digits. **(GENERAL)**

__ **Graphing Skills, CRF** This worksheet helps reinforce students' understanding of line graphs. **(GENERAL)**

__ **Concept Review, Study Guide, Movements of the Ocean** Use this worksheet to assess student's understanding of the main concepts presented in the chapter. **(GENERAL)**

__ **Concept Review, Spanish Study Guide, Movements of the Ocean** This worksheet helps reinforce English-language learners' understanding of the main concepts presented in the chapter. **(GENERAL)**

__ **Chapter Test A, CRF** Assign questions from the appropriate test for chapter assessment. **(GENERAL)**

__ **Chapter Test B, CRF** Assign questions from the appropriate test for chapter assessment. **(ADVANCED)**

__ **Spanish Chapter Test A, Spanish Assessments** Assign questions from the appropriate test for chapter assessment. **(GENERAL)**

__ **Test Generator, One-Stop Planner** Create a customized homework assignment, quiz, or test using the HRW Test Generator program. **(GENERAL)**

__ **Test Item Listing, CRF** Use the Test Item Listing to identify questions for a customized homework, quiz, or test.

__ **PowerPoint® Resources** Use the customizable presentation to help students practice standardized test-taking skills.

KEY **SE** = Student Edition **TE** = Teacher Edition **CRF** = Chapter Resource File

OTHER RESOURCE OPTIONS

__ **HRW Earth Science Video, Segment 15, "Tides"** This segment examines how the earth-moon system creates the daily rise and fall of sea level and discusses how the alignment of the sun and the moon with the Earth causes the highest tides. **(GENERAL)**

__ **NOVA Video, "Adrift on the Gulf Stream"** NOVA explores the Gulf Stream's importance to ocean life and climate from a variety of vantage points. **(GENERAL)**

__ **Brain Food Video Quiz** Use this video resource to assess student's progress. **(GENERAL)**

KEY **SE** = Student Edition **TE** = Teacher Edition **CRF** = Chapter Resource File

Chapter 22 The Atmosphere

Chapter Opener

__ **Pre-Reading Activity, Booklet, SE**

__ **Using the Figure, Storm Chasing, TE (GENERAL)**

__ **Long-Term Project, Air Pollution Watch, SE** Students collect particulates for two weeks and examine how wind direction affects their numbers. **(ADVANCED)**

__ **Long-Term Project, Air Pollution Watch, Long-Term Projects** Students use this datasheet to complete the Long-Term Project. **(ADVANCED)**

Section 1 Characteristics of the Atmosphere

PACING

Regular Schedule:	**with lab(s): N/A**	**without lab(s): 2 days**
Block Schedule:	**with lab(s): N/A**	**without lab(s): 1 day**

OBJECTIVES

1. **Describe** the composition of Earth's atmosphere.
2. **Explain** how two types of barometers work.
3. **Identify** the layers of the atmosphere.
4. **Identify** two effects of air pollution.

NATIONAL SCIENCE EDUCATION STANDARDS

PS 3c: A large number of important reactions involve the transfer of either electrons (oxidation/reduction reactions) or hydrogen ions (acid/base reactions) between reacting ions, molecules, or atoms. In other reactions, chemical bonds are broken by heat or light to form very reactive radicals with electrons ready to form new bonds. Radical reactions control many processes such as the presence of ozone and greenhouse gases in the atmosphere, burning and processing of fossil fuels, the formation of polymers, and explosions.

FOCUS *(5 minutes)*

__ **Overview, TE** Review the objectives listed in the Student Edition. **(GENERAL)**

__ **PowerPoint® Resources** Use the customizable presentation to enhance your lecture.

__ **Bellringer, TE** Students predict what will happen over time to a lighted candle covered with a glass jar. **(GENERAL)**

KEY **SE** = Student Edition **TE** = Teacher Edition **CRF** = Chapter Resource File

__ **Bellringer Transparency** Use this transparency as students enter the classroom and find their seats.

MOTIVATE *(10 minutes)*

__ **Demonstration, Air Force, TE** Use a glass bottle, a hard-boiled egg, and matches to demonstrate air pressure. **(GENERAL)**

TEACH *(20 minutes)*

__ **Directed Reading, CRF** These worksheets reinforce basic concepts and vocabulary presented in the section. **(BASIC)**

__ **Demonstration, Observing Photosynthesis, TE** Use aquatic plants to demonstrate how oxygen is produced. **(GENERAL)**

__ **Reading Skill Builder, Paired Summarizing, TE** Students read the text in pairs and take turns summarizing what they have read. **(BASIC)**

__ **Biology Connection, The Breath of Life, TE** Students research the process of respiration. **(GENERAL)**

__ **Activity, Dust Collectors, TE** Students conduct an experiment to study airborne particulates. **(BASIC)**

__ **Activity, Create a Vacuum, TE** Students use a plastic soda bottle, hot water, and ice water to create a vacuum. **(BASIC)**

__ **Quick Lab, Barometric Pressure, SE** Students build a model of a barometer that measures air density. **(GENERAL)**

__ **Datasheet for Quick Lab, Barometric Pressure, CRF** Students use the datasheet to complete the Quick Lab. **(GENERAL)**

__ **Using the Figure, Mercurial Barometers, TE** Students use the figure of a mercurial barometer to answer a question about barometric pressure. **(BASIC)**

__ **Activity, Weather Maps, TE** Students practice reading weather maps. **(GENERAL)**

__ **Teaching Transparency 106, Layers of the Atmosphere with Worksheet** Use this Teaching Transparency and the accompanying worksheet to support your classroom discussion of the layers of the atmosphere.

__ **Interactive Tutor CD-ROM, Earth's Atmosphere** Students practice what they have learned using this interactive activity. **(GENERAL)**

__ **Interactive Tutor CD-ROM, Troposphere Thickness** Students practice what they have learned using this interactive activity. **(GENERAL)**

__ **Interactive Tutor CD-ROM, Hail, Northern Lights, and Rainbows** Students practice what they have learned using this interactive activity. **(GENERAL)**

CLOSE *(10 minutes)*

__ **Section Review, SE** Students answer end-of-section vocabulary, key ideas, critical thinking, and concept mapping questions. **(GENERAL)**

KEY **SE** = Student Edition **TE** = Teacher Edition **CRF** = Chapter Resource File

__ **Homework, Ozone, Good or Bad?, TE** Students research ground-level ozone and create brochures describing the effects. **(ADVANCED)**

__ **Reteaching, Name that Sphere, TE** Students work in pairs to quiz each other about the atmosphere. **(BASIC)**

__ **Quiz, TE** Students answer 3 questions about this section. **(GENERAL)**

__ **Section Quiz, CRF** Use this quiz to assess students' understanding of the section. **(GENERAL) Also in Spanish**

__ **Spanish Section Quiz, Spanish Assessments** Use this quiz to assess students' understanding of the section. **(GENERAL)**

__ **Alternative Assessment, Modeling, TE** Students work in groups to create models of Earth's atmosphere. **(GENERAL)**

KEY **SE** = Student Edition **TE** = Teacher Edition **CRF** = Chapter Resource File

Section 2 Solar Energy and the Atmosphere

PACING

Regular Schedule:	**with lab(s): 2 days**	**without lab(s): 1 day**
Block Schedule:	**with lab(s): 1 day**	**without lab(s): 0.5 day**

OBJECTIVES

1. **Explain** how radiant energy reaches Earth.
2. **Describe** how visible light and infrared energy warm Earth.
3. **Summarize** the processes of radiation, conduction, and convection.

NATIONAL SCIENCE EDUCATION STANDARDS

ES 1c: Heating of earth's surface and atmosphere by the sun drives convection within the atmosphere and oceans, producing winds and ocean currents.

ES 1d: Global climate is determined by energy transfer from the sun at and near the earth's surface. This energy transfer is influenced by dynamic processes such as cloud cover and the earth's rotation, and static conditions such as the position of mountain ranges and oceans.

PS 3c: A large number of important reactions involve the transfer of either electrons (oxidation/reduction reactions) or hydrogen ions (acid/base reactions) between reacting ions, molecules, or atoms. In other reactions, chemical bonds are broken by heat or light to form very reactive radicals with electrons ready to form new bonds. Radical reactions control many processes such as the presence of ozone and greenhouse gases in the atmosphere, burning and processing of fossil fuels, the formation of polymers, and explosions.

PS 6a: Waves, including sound and seismic waves, waves on water, and light waves, have energy and can transfer energy when they interact with matter.

PS 6b: Electromagnetic waves result when a charged object is accelerated or decelerated. Electromagnetic waves include radio waves (the longest wavelength), microwaves, infrared radiation (radiant heat), visible light, ultraviolet radiation, x-rays, and gamma rays. The energy of electromagnetic waves is carried in packets whose magnitude is inversely proportional to the wavelength.

KEY **SE** = Student Edition **TE** = Teacher Edition **CRF** = Chapter Resource File

FOCUS *(5 minutes)*

__ **Overview, TE** Review the objectives listed in the Student Edition. **(GENERAL)**

__ **PowerPoint® Resources** Use the customizable presentation to enhance your lecture.

__ **Bellringer, TE** Ask students to describe what effect the sun has on Earth. **(GENERAL)**

__ **Bellringer Transparency** Use this transparency as students enter the classroom and find their seats.

MOTIVATE *(10 minutes)*

__ **Activity, Magic with Beads, TE** Use ultraviolet sensing beads to demonstrate how sunlight differs from indoor light. **(GENERAL)**

TEACH *(65 minutes)*

__ **Directed Reading, CRF** These worksheets reinforce basic concepts and vocabulary presented in the section. **(BASIC)**

__ **Activity, Blue and Red Skies, TE** Students use a glass, water, milk, and a flashlight to model how light travels through the atmosphere. **(GENERAL)**

__ **Skill Builder, Writing, TE** Students research and write essays about chemical alternatives to CFCs. **(ADVANCED)**

__ **Group Activity, Comparing Albedos, TE** Students conduct an experiment comparing albedos. **(GENERAL)**

__ **Using the Figure, Phantom Images, TE** Students use the photo of a desert to answer a question about mirages. **(GENERAL)**

__ **Skill Builder, Vocabulary, TE** Students review the origin of the word *albedo*. **(BASIC)**

__ **Internet Activity, Global Warming, TE** Students research the causes and potential effects of an enhanced greenhouse effect on Earth. **(GENERAL)**

__ **Internet Activity, Global Warming, CRF** This worksheet directs student research for the topic presented in the Teacher Edition. **(GENERAL)**

__ **Teaching Transparency 108, The Greenhouse Effect and Latitude and Season with Worksheet** Use this Teaching Transparency and the accompanying worksheet to support your classroom discussion of the greenhouse effect and latitude and season.

__ **Quick Lab, Light and Latitude, SE** Use a flashlight, a white piece of paper, and a pencil to demonstrate how latitude affects incoming solar radiation. **(GENERAL)**

__ **Datasheet for Quick Lab, Light and Latitude, CRF** Students use the datasheet to complete the Quick Lab. **(GENERAL)**

__ **Inquiry Lab, Energy Absorption and Reflection, SE** Students perform an experiment to observe how energy is absorbed and reflected. **(GENERAL)**

__ **Datasheet for Chapter Lab, Inquiry Lab, Energy Absorption and Reflection, CRF** Students use the datasheet to complete the Chapter Lab. **(GENERAL)**

KEY **SE** = Student Edition **TE** = Teacher Edition **CRF** = Chapter Resource File

__ **Maps in Action, Absorbed Solar Radiation, SE** Students enhance their map-reading skills by answering questions about a graphic map showing the total amount of solar radiation that Earth absorbs in January and July. **(GENERAL)**

__ **Teaching Transparency 111, Absorbed Solar Radiation with Worksheet** Use this Teaching Transparency and the accompanying worksheet to help students perform the Maps in Action activity.

__ **Discussion, The Tropics** Students use the map of absorbed solar radiation to answer questions about the tropics. **(GENERAL)**

__ **Inquiry Lab, Ultraviolet Protection, CRF** Students use UV-sensitive paper to measure the effectiveness of sunscreen and fabrics against UV radiation. **(GENERAL)**

__ **Interactive Tutor CD-ROM, Greenhouse Effect** Students practice what they have learned using this interactive activity. **(GENERAL)**

__ **Interactive Tutor CD-ROM, Energy** Students practice what they have learned using this interactive activity. **(GENERAL)**

__ **Interactive Tutor CD-ROM, Light and Electromagnetic Energy** Students practice what they have learned using this interactive activity. **(GENERAL)**

CLOSE *(10 minutes)*

__ **Section Review, SE** Students answer end-of-section vocabulary, key ideas, critical thinking, and concept mapping questions. **(GENERAL)**

__ **Reteaching, Peer Reviewing, TE** Students work in pairs to quiz each other about the material in this section. **(BASIC)**

__ **Quiz, TE** Students answer 3 questions about this section. **(GENERAL)**

__ **Section Quiz, CRF** Use this quiz to assess students' understanding of the section. **(GENERAL) Also in Spanish**

__ **Spanish Section Quiz, Spanish Assessments** Use this quiz to assess students' understanding of the section. **(GENERAL)**

__ **Alternative Assessment, The Spectrum of Life, TE** Students keep a journal for one week about electromagnetic radiation. **(GENERAL)**

KEY SE = Student Edition **TE** = Teacher Edition **CRF** = Chapter Resource File

Section 3 Atmospheric Circulation

PACING

Regular Schedule: **with lab(s): N/A** **without lab(s): 2 days**
Block Schedule: **with lab(s): N/A** **without lab(s): 1 day**

OBJECTIVES

1. **Explain** the Coriolis effect.
2. **Describe** the global patterns of air circulation, and name three global wind belts.
3. **Identify** two factors that form local wind patterns.

NATIONAL SCIENCE EDUCATION STANDARDS

ES 1c: Heating of earth's surface and atmosphere by the sun drives convection within the atmosphere and oceans, producing winds and ocean currents.

ES 1d: Global climate is determined by energy transfer from the sun at and near the earth's surface. This energy transfer is influenced by dynamic processes such as cloud cover and the earth's rotation, and static conditions such as the position of mountain ranges and oceans.

FOCUS *(5 minutes)*

__ **Overview, TE** Review the objectives listed in the Student Edition. **(GENERAL)**

__ **PowerPoint® Resources** Use the customizable presentation to enhance your lecture.

__ **Bellringer, TE** Ask students to describe what causes air to move. **(GENERAL)**

__ **Bellringer Transparency** Use this transparency as students enter the classroom and find their seats.

MOTIVATE *(10 minutes)*

__ **Demonstration, Modeling the Coriolis Effect** Use a paper plate, a turntable, and a pencil to demonstrate the Coriolis effect. **(GENERAL)**

TEACH *(20 minutes)*

__ **Directed Reading, CRF** These worksheets reinforce basic concepts and vocabulary presented in the section. **(BASIC)**

__ **Teaching Transparency 109, The Coriolis Effect with Worksheet** Use this Teaching Transparency and the accompanying worksheet to support your classroom discussion of the Coriolis effect.

KEY **SE** = Student Edition **TE** = Teacher Edition **CRF** = Chapter Resource File

__ **Using the Figure, Coriolis Effect, TE** Students use the figure of the hemispheres to answer a question about winds in the Northern Hemisphere. **(ADVANCED)**

__ **Group Activity, It's a Breeze, TE** Students conduct an experiment that creates a breeze. **(GENERAL)**

__ **Teaching Transparency 110, Global Wind Belts with Worksheet** Use this Teaching Transparency and the accompanying worksheet to support your classroom discussion of global wind belts.

__ **Reading Skill Builder, Reading Organizer, TE** Students are encouraged to take notes while reading this section. **(BASIC)**

__ **Graphic Organizer, Comparison Table, SE** Use the graphic organizer to help students develop their analytical skills. **(GENERAL)**

__ **Making Models Lab, Global Air Movement, CRF** Students collect particulates for two weeks and examine how wind direction affects their numbers. **(GENERAL)**

__ **Science and Technology, Energy from the Wind, SE** The Darrieus turbine and the use of wind as an alternative energy source is discussed in this article.

__ **Internet Activity, Harnessing Wind, TE** Students research the benefits and drawbacks of using wind energy as an alternative energy source. **(GENERAL)**

__ **Internet Activity, Harnessing Wind, CRF** This worksheet directs student research for the topic presented in the Teacher Edition. **(GENERAL)**

__ **Interactive Tutor CD-ROM, Global Wind Patterns** Students practice what they have learned using this interactive activity. **(GENERAL)**

CLOSE *(10 minutes)*

__ **Section Review, SE** Students answer end-of-section vocabulary, key ideas, critical thinking, and concept mapping questions. **(GENERAL)**

__ **Reteaching, Labeling, TE** Students use a map or globe to identify the locations and prevailing directions of the major wind belts. **(BASIC)**

__ **Quiz, TE** Students answer 3 questions about this section. **(GENERAL)**

__ **Section Quiz, CRF** Use this quiz to assess students' understanding of the section. **(GENERAL) Also in Spanish**

__ **Spanish Section Quiz, Spanish Assessments** Use this quiz to assess students' understanding of the section. **(GENERAL)**

__ **Alternative Assessment, TE** Students write essays explaining why wind energy is considered as a form of solar energy. **(ADVANCED)**

KEY **SE** = Student Edition **TE** = Teacher Edition **CRF** = Chapter Resource File

End of Chapter Review and Assessment

PACING

Regular Schedule: **with lab(s): N/A** **without lab(s): 2 days**
Block Schedule: **with lab(s): N/A** **without lab(s): 1 day**

__ **Chapter Review, SE** Students answer end-of-chapter vocabulary, key ideas, critical thinking, math, writing, and interpreting graphics questions. **(GENERAL)**

__ **Alternative Assessment, Life's a Gas, TE** Students imagine that they are an atmospheric gas and write stories describing their lives. **(ADVANCED)**

__ **Standardized Test Prep, SE** Have students complete the standardized test prep to help the prepare for standardized tests. **(GENERAL)**

__ **Holt PuzzlePro®** Use this software to create crossword puzzles and word searches that make learning vocabulary fun. **(GENERAL)**

__ **Concept Review, CRF** Use this worksheet to assess student's understanding of the main concepts presented in the chapter. **(GENERAL)**

__ **Critical Thinking, CRF** Students use this worksheet to develop and practice critical thinking skills. **(ADVANCED)**

__ **Math Skills, CRF** This worksheet helps reinforce students' understanding of significant digits. **(GENERAL)**

__ **Graphing Skills, CRF** This worksheet helps reinforce students' understanding of line graphs. **(GENERAL)**

__ **Concept Review, Study Guide, The Atmosphere** Use this worksheet to assess student's understanding of the main concepts presented in the chapter. **(GENERAL)**

__ **Concept Review, Spanish Study Guide, The Atmosphere** This worksheet helps reinforce English-language learners' understanding of the main concepts presented in the chapter. **(GENERAL)**

__ **Chapter Test A, CRF** Assign questions from the appropriate test for chapter assessment. **(GENERAL)**

__ **Chapter Test B, CRF** Assign questions from the appropriate test for chapter assessment. **(ADVANCED)**

__ **Spanish Chapter Test A, Spanish Assessments** Assign questions from the appropriate test for chapter assessment. **(GENERAL)**

__ **Test Generator, One-Stop Planner** Create a customized homework assignment, quiz, or test using the HRW Test Generator program. **(GENERAL)**

__ **Test Item Listing, CRF** Use the Test Item Listing to identify questions for a customized homework, quiz, or test.

__ **PowerPoint® Resources** Use the customizable presentation to help students practice standardized test-taking skills.

KEY **SE** = Student Edition **TE** = Teacher Edition **CRF** = Chapter Resource File

OTHER RESOURCE OPTIONS

__ **HRW Earth Science Video, Segment 16, "Global Winds"** This segment focuses on winds that blow steadily over large areas of the planet's surface. The existence of pressure belts at different latitudes is examined, and the convection cells that are created by these differences in pressure are discussed. The Coriolis effect is also described, and global wind patterns are shown. **(GENERAL)**

__ **HRW Earth Science Video, Segment 17, "Local Winds"** In this segment, the relative rates of warming and cooling over land and water and the ascending and descending air masses that result in land and sea breezes are discussed. Also explored are local temperature differences in mountainous areas that result in mountain and valley breezes. **(GENERAL)**

__ **Brain Food Video Quiz** Use this video resource to assess student's progress. **(GENERAL)**

KEY **SE** = Student Edition **TE** = Teacher Edition **CRF** = Chapter Resource File

Chapter 23 Water in the Atmosphere

Chapter Opener

__ **Pre-Reading Activity, Two-Panel Flip Chart, SE**

__ **Using the Figure, Clouds and Weather, TE (GENERAL)**

Section 1 Atmospheric Moisture

PACING

Regular Schedule:	**with lab(s): 2 days**	**without lab(s): 1 day**
Block Schedule:	**with lab(s): 1 day**	**without lab(s): 0.5 day**

OBJECTIVES

1. **Explain** how heat energy affects the changing phases of water.
2. **Explain** what absolute humidity and relative humidity are, and describe how they are measured.
3. **Describe** what happens when the temperature of air decreases to the dew point or below the dew point.

NATIONAL SCIENCE EDUCATION STANDARDS

ES 1c: Heating of earth's surface and atmosphere by the sun drives convection within the atmosphere and oceans, producing winds and ocean currents.

ES 1d: Global climate is determined by energy transfer from the sun at and near the earth's surface. This energy transfer is influenced by dynamic processes such as cloud cover and the earth's rotation, and static conditions such as the position of mountain ranges and oceans.

ES 2b: Movement of matter between reservoirs is driven by the earth's internal and external sources of energy. These movements are often accompanied by a change in the physical and chemical properties of the matter. Carbon, for example, occurs in carbonate rocks such as limestone, in the atmosphere as carbon dioxide gas, in water as dissolved carbon dioxide, and in all organisms as complex molecules that control the chemistry of life.

FOCUS *(5 minutes)*

__ **Overview, TE** Review the objectives listed in the Student Edition. **(GENERAL)**

__ **PowerPoint® Resources** Use the customizable presentation to enhance your lecture.

KEY **SE** = Student Edition **TE** = Teacher Edition **CRF** = Chapter Resource File

__ **Bellringer, TE** Ask students to describe what must happen to ice to cause it to melt and what must happen to water to cause it to boil. **(GENERAL)**

__ **Bellringer Transparency** Use this transparency as students enter the classroom and find their seats.

MOTIVATE *(10 minutes)*

__ **Activity, Cooling Effect, TE** Students rub alcohol swabs on their skin to illustrate how quickly alcohol evaporates. **(GENERAL)**

TEACH *(65 minutes)*

__ **Directed Reading, CRF** These worksheets reinforce basic concepts and vocabulary presented in the section. **(BASIC)**

__ **Teaching Transparency 112, Phases of Water with Worksheet** Use this Teaching Transparency and the accompanying worksheet to support your classroom discussion of the phases of water.

__ **Physics Connection, Atmospheric Effects of Light, TE** Students research sun dogs, mirages, and glories and present their findings. **(ADVANCED)**

__ **Reading Skill Builder, Discussion, TE** Students discuss how relative humidity relates to physical comfort. **(BASIC)**

__ **Teaching Transparency 113, Vapor Pressure with Worksheet** Use this Teaching Transparency and the accompanying worksheet to support your classroom discussion of vapor pressure.

__ **Biology Connection, Desert Survival, TE** Students research sun dogs, mirages, and glories and present their findings. **(ADVANCED)**

__ **Activity, Scientific Instruments, TE** Students study how real meteorological instruments work. **(GENERAL)**

__ **Quick Lab, Dew Point, SE** Students use a glass container, water, and ice to conduct an experiment about condensation. **(GENERAL)**

__ **Datasheet for Quick Lab, Dew Point, CRF** Students use the datasheet to complete the Quick Lab. **(GENERAL)**

__ **Skills Practice Lab, Relative Humidity, SE** Students use wet-bulb and dry-bulb thermometer readings to determine the relative humidity of the air in the classroom. **(GENERAL)**

__ **Datasheet for Chapter Lab, Skills Practice Lab, Relative Humidity, CRF** Students use the datasheet to complete the Chapter Lab. **(GENERAL)**

__ **Mapping Expeditions, What Comes Down Must Go…Where? SE** Students observe where rainwater goes after it rains and identify sources of pollution. **(GENERAL)**

__ **Interactive Tutor CD-ROM, Ice, Water, and Vapor** Students practice what they have learned using this interactive activity. **(GENERAL)**

__ **Interactive Tutor CD-ROM, The Water Cycle** Students practice what they have learned using this interactive activity. **(GENERAL)**

KEY SE = Student Edition **TE** = Teacher Edition **CRF** = Chapter Resource File

CLOSE *(10 minutes)*

__ **Section Review, SE** Students answer end-of-section vocabulary, key ideas, critical thinking, and concept mapping questions. **(GENERAL)**

__ **Reteaching, Making Diagrams, TE** Students work in pairs to create diagrams illustrating chapter concepts. **(BASIC)**

__ **Quiz, TE** Students answer 4 questions about this section. **(GENERAL)**

__ **Section Quiz, CRF** Use this quiz to assess students' understanding of the section. **(GENERAL) Also in Spanish**

__ **Spanish Section Quiz, Spanish Assessments** Use this quiz to assess students' understanding of the section. **(GENERAL)**

__ **Alternative Assessment, Song Lyrics, TE** Students compose a song about energy and phase changes. **(ADVANCED)**

KEY **SE** = Student Edition **TE** = Teacher Edition **CRF** = Chapter Resource File

Lesson Plan

Section 2 Clouds and Fog

PACING

Regular Schedule:	**with lab(s): N/A**	**without lab(s): 2 days**
Block Schedule:	**with lab(s): N/A**	**without lab(s): 1 day**

OBJECTIVES

1. **Describe** the conditions that are necessary for clouds to form.
2. **Explain** the four processes of cooling that can lead to the formation of clouds.
3. **Identify** the three types of clouds.
4. **Describe** four ways in which fog can form.

NATIONAL SCIENCE EDUCATION STANDARDS

ES 1c: Heating of earth's surface and atmosphere by the sun drives convection within the atmosphere and oceans, producing winds and ocean currents.

ES 1d: Global climate is determined by energy transfer from the sun at and near the earth's surface. This energy transfer is influenced by dynamic processes such as cloud cover and the earth's rotation, and static conditions such as the position of mountain ranges and oceans.

ES 2b: Movement of matter between reservoirs is driven by the earth's internal and external sources of energy. These movements are often accompanied by a change in the physical and chemical properties of the matter. Carbon, for example, occurs in carbonate rocks such as limestone, in the atmosphere as carbon dioxide gas, in water as dissolved carbon dioxide, and in all organisms as complex molecules that control the chemistry of life.

PS 2d: The physical properties of compounds reflect the nature of the interactions among its molecules. These interactions are determined by the structure of the molecule, including the constituent atoms and the distances and angles between them.

FOCUS *(5 minutes)*

__ **Overview, TE** Review the objectives listed in the Student Edition. **(GENERAL)**

__ **PowerPoint® Resources** Use the customizable presentation to enhance your lecture.

__ **Bellringer, TE** Ask students to think about kinds of clouds they have seen and write descriptions of the different cloud formations. **(GENERAL)**

KEY **SE** = Student Edition **TE** = Teacher Edition **CRF** = Chapter Resource File

Lesson Plan

__ **Bellringer Transparency** Use this transparency as students enter the classroom and find their seats.

MOTIVATE *(10 minutes)*

__ **Demonstration, Condensation Nuclei, TE** Use salt, water, a microscope slide, and a petri dish to demonstrate evaporation and condensation. **(GENERAL)**

TEACH *(20 minutes)*

__ **Directed Reading, CRF** These worksheets reinforce basic concepts and vocabulary presented in the section. **(BASIC)**

__ **Teaching Transparency 114, Formation of a Water Droplet with Worksheet** Use this Teaching Transparency and the accompanying worksheet to support your classroom discussion of the formation of a water droplet.

__ **Internet Activity, Latent Heat and Thunderstorms, TE** Students research latent heat in exceptional weather patterns. **(GENERAL)**

__ **Internet Activity, Latent Heat and Thunderstorms, CRF** This worksheet directs student research for the topic presented in the Teacher Edition. **(GENERAL)**

__ **Using the Figure, Discussion, TE** Students use the figure showing dew point temperature versus cloud height to answer a question about dew point. **(BASIC)**

__ **Math Skill Builder, Adiabatic Lapse Rate, TE** Students calculate how much the dew point would drop for given elevation heights. **(GENERAL)**

__ **Reading Skill Builder, Paired Summarizing, TE** Students read the text in pairs and take turns summarizing what they have read. **(BASIC)**

__ **Geography Connection, Rain Shadow, TE** Students research the rain shadow effect and find examples of this phenomenon. **(GENERAL)**

__ **Quick Lab, Cloud Formation, SE** Students perform an experiment about latent heat. **(GENERAL)**

__ **Datasheet for Quick Lab, Cloud Formation, CRF** Students use the datasheet to complete the Quick Lab. **(GENERAL)**

__ **Teaching Transparency 115, Classification of Clouds with Worksheet** Use this Teaching Transparency and the accompanying worksheet to support your classroom discussion of the classification of clouds.

__ **Using the Figure, Cloud Bulletin Board, TE** Students use the figure showing different cloud types to create a bulletin board illustrating the different cloud types. **(GENERAL)**

__ **Skill Builder, Vocabulary, TE** Students review the origin of the names for the three basic cloud shapes. **(GENERAL)**

__ **Graphic Organizer, Comparison Table, SE** Use the graphic organizer to help students develop their analytical skills. **(GENERAL)**

__ **Interactive Tutor CD-ROM, Where Do Clouds Form?** Students practice what they have learned using this interactive activity. **(GENERAL)**

KEY **SE** = Student Edition **TE** = Teacher Edition **CRF** = Chapter Resource File

Lesson Plan

CLOSE *(10 minutes)*

__ **Section Review, SE** Students answer end-of-section vocabulary, key ideas, critical thinking, and concept mapping questions. **(GENERAL)**

__ **Reteaching, Cloud Lingo, TE** Students work in pairs to create index cards with key terms and definitions from the section. **(BASIC)**

__ **Quiz, TE** Students answer 2 questions about this section. **(GENERAL)**

__ **Section Quiz, CRF** Use this quiz to assess students' understanding of the section. **(GENERAL) Also in Spanish**

__ **Spanish Section Quiz, Spanish Assessments** Use this quiz to assess students' understanding of the section. **(GENERAL)**

__ **Alternative Assessment, Cloud Atlas, TE** Students create a Cloud Atlas based on the modern cloud classification system. **(ADVANCED)**

KEY **SE** = Student Edition **TE** = Teacher Edition **CRF** = Chapter Resource File

Lesson Plan

Section 3 Precipitation

PACING

Regular Schedule: **with lab(s): N/A** **without lab(s): 2 days**
Block Schedule: **with lab(s): N/A** **without lab(s): 1 day**

OBJECTIVES

1. **Identify** the four forms of precipitation.
2. **Compare** the two processes that cause precipitation.
3. **Describe** two ways that precipitation is measured.
4. **Explain** how rain can be produced artificially.

NATIONAL SCIENCE EDUCATION STANDARDS

ES 1c: Heating of earth's surface and atmosphere by the sun drives convection within the atmosphere and oceans, producing winds and ocean currents.

ES 2b: Movement of matter between reservoirs is driven by the earth's internal and external sources of energy. These movements are often accompanied by a change in the physical and chemical properties of the matter. Carbon, for example, occurs in carbonate rocks such as limestone, in the atmosphere as carbon dioxide gas, in water as dissolved carbon dioxide, and in all organisms as complex molecules that control the chemistry of life.

LS 4e: Human beings live within the world's ecosystems. Increasingly, humans modify ecosystems as a result of population growth, technology, and consumption. Human destruction of habitats through direct harvesting, pollution, atmospheric changes, and other factors is threatening current global stability, and if not addressed, ecosystems will be irreversibly affected.

FOCUS *(5 minutes)*

__ **Overview, TE** Review the objectives listed in the Student Edition. **(GENERAL)**

__ **PowerPoint® Resources** Use the customizable presentation to enhance your lecture.

__ **Bellringer, TE** Ask students why they think the water droplets that form clouds often fall to Earth as rain or other forms of precipitation. **(GENERAL)**

__ **Bellringer Transparency** Use this transparency as students enter the classroom and find their seats.

KEY **SE** = Student Edition **TE** = Teacher Edition **CRF** = Chapter Resource File

Lesson Plan

MOTIVATE *(10 minutes)*

__ **Demonstration, Indoor Rain, TE** Use a hot plate, a jar, ice cubes, a zip-top plastic bag and aluminum foil to demonstrate evaporation and condensation. **(GENERAL)**

TEACH *(20 minutes)*

__ **Directed Reading, CRF** These worksheets reinforce basic concepts and vocabulary presented in the section. **(BASIC)**

__ **Group Activity, Supercooled: Test Tube Hail, TE** Students use beakers, clean test tubes, thermometers, crushed ice, and salt to create hail. **(GENERAL)**

__ **Maps in Action, Annual Precipitation in the United States, SE** Students enhance their map-reading skills by answering questions about map showing the total precipitation for the year 2003 for different regions of the United States. **(GENERAL)**

__ **Teaching Transparency 116, Annual Precipitation in the United States with Worksheet** Use this Teaching Transparency and the accompanying worksheet to help students perform the Maps in Action activity.

__ **Internet Activity, Climate and Precipitation, TE** Students research the annual rates of two forms of precipitation for one region of the United States. **(GENERAL)**

__ **Internet Activity, Climate and Precipitation, CRF** This worksheet directs student research for the topic presented in the Teacher Edition. **(GENERAL)**

__ **Impact on Society, Hail, SE** The formation and the destructive power of hail is discussed.

__ **Activity, Research, TE** Students research hail. **(GENERAL)**

__ **Inquiry Lab, How Big Is a Raindrop? CRF** Students use real raindrops to determine the size of raindrops. **(GENERAL)**

__ **Making Models Lab, What Is the Shape of a Raindrop? CRF** Students model the way raindrops behave in a cloud to determine the shape of raindrops. **(GENERAL)**

__ **Interactive Tutor CD-ROM, Hail, Northern Lights, and Rainbows** Students practice what they have learned using this interactive activity. **(GENERAL)**

CLOSE *(10 minutes)*

__ **Section Review, SE** Students answer end-of-section vocabulary, key ideas, critical thinking, and concept mapping questions. **(GENERAL)**

__ **Homework, Measuring Rainfall, TE** Students build a simple rain gauge. **(GENERAL)**

__ **Reteaching, Comparison Table, TE** Students work in pairs and compare the different forms of precipitation. **(BASIC)**

__ **Quiz, TE** Students answer 3 questions about this section. **(GENERAL)**

__ **Section Quiz, CRF** Use this quiz to assess students' understanding of the section. **(GENERAL) Also in Spanish**

KEY SE = Student Edition TE = Teacher Edition CRF = Chapter Resource File

__ **Spanish Section Quiz, Spanish Assessments** Use this quiz to assess students' understanding of the section. **(GENERAL)**

__ **Alternative Assessment, Instruction Manuals, TE** Students write an instruction manual explaining how different types of precipitation are formed in the atmosphere. **(GENERAL)**

KEY SE = Student Edition **TE** = Teacher Edition **CRF** = Chapter Resource File

End of Chapter Review and Assessment

PACING

Regular Schedule:	**with lab(s): N/A**	**without lab(s): 2 days**
Block Schedule:	**with lab(s): N/A**	**without lab(s): 1 day**

__ **Chapter Review, SE** Students answer end-of-chapter vocabulary, key ideas, critical thinking, math, writing, and interpreting graphics questions. **(GENERAL)**

__ **Alternative Assessment, Water Cycle Skit, TE** Students perform skits explaining how water moves through the water cycle. **(GENERAL)**

__ **Standardized Test Prep, SE** Have students complete the standardized test prep to help them prepare for standardized tests. **(GENERAL)**

__ **Holt PuzzlePro®** Use this software to create crossword puzzles and word searches that make learning vocabulary fun. **(GENERAL)**

__ **Concept Review, CRF** Use this worksheet to assess student's understanding of the main concepts presented in the chapter. **(GENERAL)**

__ **Critical Thinking, CRF** Students use this worksheet to develop and practice critical thinking skills. **(ADVANCED)**

__ **Math Skills, CRF** This worksheet helps reinforce students' understanding of algebraic rearrangements. **(GENERAL)**

__ **Graphing Skills, CRF** This worksheet helps reinforce students' understanding of pie graphs. **(GENERAL)**

__ **Concept Review, Study Guide, Water in the Atmosphere** Use this worksheet to assess student's understanding of the main concepts presented in the chapter. **(GENERAL)**

__ **Concept Review, Spanish Study Guide, Water in the Atmosphere** This worksheet helps reinforce English-language learners' understanding of the main concepts presented in the chapter. **(GENERAL)**

__ **Chapter Test A, CRF** Assign questions from the appropriate test for chapter assessment. **(GENERAL)**

__ **Chapter Test B, CRF** Assign questions from the appropriate test for chapter assessment. **(ADVANCED)**

__ **Spanish Chapter Test A, Spanish Assessments** Assign questions from the appropriate test for chapter assessment. **(GENERAL)**

__ **Test Generator, One-Stop Planner** Create a customized homework assignment, quiz, or test using the HRW Test Generator program. **(GENERAL)**

__ **Test Item Listing, CRF** Use the Test Item Listing to identify questions for a customized homework, quiz, or test.

__ **PowerPoint® Resources** Use the customizable presentation to help students practice standardized test-taking skills.

KEY **SE** = Student Edition **TE** = Teacher Edition **CRF** = Chapter Resource File

OTHER RESOURCE OPTIONS

_ **Brain Food Video Quiz** Use this video resource to assess student's progress. **(GENERAL)**

KEY **SE** = Student Edition **TE** = Teacher Edition **CRF** = Chapter Resource File

Chapter 24 Weather

Chapter Opener

__ **Pre-Reading Activity, Key Term Fold, SE**

__ **Using the Figure, Light Show, TE (GENERAL)**

__ **Long-Term Project, Correlating Weather Variables, SE** Students collect weather data for one month, organize and graph it, and make predictions. **(ADVANCED)**

__ **Long-Term Project, Correlating Weather Variables, Long-term Projects** Students use this datasheet to complete the Long-Term Project. **(ADVANCED)**

__ **Long-Term Project, Weather Forecasting, SE** Students use a series of weather maps to track weather system movements for one week. **(ADVANCED)**

__ **Long-Term Project, Weather Forecasting, Long-Term Projects** Students use this datasheet to complete the Long-term Project. **(ADVANCED)**

Section 1 Air Masses

PACING

Regular Schedule:	**with lab(s): N/A**	**without lab(s): 2 days**
Block Schedule:	**with lab(s): N/A**	**without lab(s): 1 day**

OBJECTIVES

1. **Explain** how an air mass forms.
2. **List** the four main types of air masses.
3. **Describe** how air masses affect the weather of North America.

NATIONAL SCIENCE EDUCATION STANDARDS

HNS 2a: Science distinguishes itself from other ways of knowing and from other bodies of knowledge through the use of empirical standards, logical arguments, and skepticism, as scientists strive for the best possible explanations about the natural world.

SAI 2a: Scientists usually inquire about how physical, living, or designed systems function. Conceptual principles and knowledge guide scientific inquiries. Historical and current scientific knowledge influence the design and interpretation of investigations and the evaluation of proposed explanations made by other scientists.

UCP 2: Evidence consists of observations and data on which to base scientific explanations. Using evidence to understand interactions allows individuals to predict changes in natural and designed systems. Models are tentative schemes or structures that correspond to real objects, events, or classes of events, and that have explanatory power. Models help scientists and engineers understand how things work. Models take many forms, including physical objects, plans, mental constructs, mathematical equations, and computer simulations.

KEY **SE** = Student Edition **TE** = Teacher Edition **CRF** = Chapter Resource File

Lesson Plan

UCP 3: Although most things are in the process of becoming different—changing—some properties of objects and processes are characterized by constancy, including the speed of light, the charge of an electron, and the total mass plus energy in the universe. Changes might occur, for example, in properties of materials, position of objects, motion, and form and function of systems. Interactions within and among systems result in change. Changes vary in rate, scale, and pattern, including trends and cycles.

FOCUS *(5 minutes)*

__ **Overview, TE** Review the objectives listed in the Student Edition. **(GENERAL)**

__ **PowerPoint® Resources** Use the customizable presentation to enhance your lecture.

__ **Bellringer, TE** Students compare what they know of land and ocean air and polar and tropical air. **(GENERAL)**

__ **Bellringer Transparency** Use this transparency as students enter the classroom and find their seats.

MOTIVATE *(10 minutes)*

__ **Demonstration, Coriolis Effect, TE** Use a globe to demonstrate how wind patterns curve. **(GENERAL)**

TEACH *(20 minutes)*

__ **Directed Reading, CRF** These worksheets reinforce basic concepts and vocabulary presented in the section. **(BASIC)**

__ **Demonstration, Comparing Air Masses, TE** Students explain how stable air masses form and compare characteristics of different air masses. **(GENERAL)**

__ **Activity, Air on the Move, TE** Students identify sources of air masses. **(GENERAL)**

__ **Reading Skill Builder, Reading Hint, TE** Students discuss the meanings of adjectives used to describe air masses to better understand how they move. **(BASIC)**

__ **Activity, Bulletin Board Project, TE** Students create a weather bulletin board using information on North American air masses. **(GENERAL)**

__ **Teaching Transparency 117, Air Masses with Worksheet** Use this Teaching Transparency and the accompanying worksheet to support your classroom discussion of air masses.

__ **Interactive Tutor CD-ROM, North American Weather** Students practice what they have learned using this interactive activity. **(GENERAL)**

CLOSE *(10 minutes)*

__ **Section Review, SE** Students answer end-of-section vocabulary, key ideas, critical thinking, and concept mapping questions. **(GENERAL)**

__ **Reteaching, Spider Maps, TE** Students create spider maps describing air masses. **(BASIC)**

__ **Quiz, TE** Students answer 4 questions about this section. **(GENERAL)**

KEY **SE** = Student Edition **TE** = Teacher Edition **CRF** = Chapter Resource File

Lesson Plan

__ **Section Quiz, CRF** Use this quiz to assess students' understanding of the section. **(GENERAL) Also in Spanish**

__ **Spanish Section Quiz, Spanish Assessments** Use this quiz to assess students' understanding of the section. **(GENERAL)**

__ **Alternative Assessment, Electric Quiz Game, TE** Students make a matching game using the characteristics of air masses. **(ADVANCED)**

KEY **SE** = Student Edition **TE** = Teacher Edition **CRF** = Chapter Resource File

Section 2 Fronts

PACING

Regular Schedule: **with lab(s): N/A** **without lab(s): 2 days**
Block Schedule: **with lab(s): N/A** **without lab(s): 1 day**

OBJECTIVES

1. **Compare** the characteristic weather patterns of cold fronts with those of warm fronts.
2. **Describe** how a midlatitude cyclone forms.
3. **Describe** the development of hurricanes, thunderstorms, and tornadoes.

NATIONAL SCIENCE EDUCATION STANDARDS

HNS 2a: Science distinguishes itself from other ways of knowing and from other bodies of knowledge through the use of empirical standards, logical arguments, and skepticism, as scientists strive for the best possible explanations about the natural world.

SAI 2a: Scientists usually inquire about how physical, living, or designed systems function. Conceptual principles and knowledge guide scientific inquiries. Historical and current scientific knowledge influence the design and interpretation of investigations and the evaluation of proposed explanations made by other scientists.

UCP 2: Evidence consists of observations and data on which to base scientific explanations. Using evidence to understand interactions allows individuals to predict changes in natural and designed systems. Models are tentative schemes or structures that correspond to real objects, events, or classes of events, and that have explanatory power. Models help scientists and engineers understand how things work. Models take many forms, including physical objects, plans, mental constructs, mathematical equations, and computer simulations.

UCP 3: Although most things are in the process of becoming different—changing—some properties of objects and processes are characterized by constancy, including the speed of light, the charge of an electron, and the total mass plus energy in the universe. Changes might occur, for example, in properties of materials, position of objects, motion, and form and function of systems. Interactions within and among systems result in change. Changes vary in rate, scale, and pattern, including trends and cycles.

FOCUS *(5 minutes)*

__ **Overview, TE** Review the objectives listed in the Student Edition. (**GENERAL**)

__ **PowerPoint® Resources** Use the customizable presentation to enhance your lecture.

__ **Bellringer, TE** Students select a weather condition and describe how it affects people. (**GENERAL**)

__ **Bellringer Transparency** Use this transparency as students enter the classroom and find their seats.

KEY **SE** = Student Edition **TE** = Teacher Edition **CRF** = Chapter Resource File

MOTIVATE *(10 minutes)*

__ **Demonstration, Make a Cold Front, TE** Students observe the effects of meeting air masses. **(GENERAL)**

TEACH *(20 minutes)*

__ **Directed Reading, CRF** These worksheets reinforce basic concepts and vocabulary presented in the section. **(BASIC)**

__ **Teaching Transparency 118, Types of Fronts with Worksheet** Use this Teaching Transparency and the accompanying worksheet to support your classroom discussion of types of fronts.

__ **Group Activity, Weather Front Pop-Ups, TE** Students create booklets showing what happens when different types of fronts form. **(GENERAL)**

__ **Reading Skill Builder, Paired Summarizing, TE** Students read the text in pairs and take turns summarizing what they have read. **(BASIC)**

__ **History Connection, Pioneer Meteorologist, TE** Students research the work done by Vilhelm and Jacob Bjerknes. **(ADVANCED)**

__ **Skill Builder, Vocabulary, TE** Students discuss the prefix *anti-* and how it relates to *cyclone* and other science words. **(BASIC)**

__ **Teaching Transparency 119, Stages of a Midlatitude Cyclone with Worksheet** Use this Teaching Transparency and the accompanying worksheet to support your classroom discussion of the stages of a midlatitude cyclone.

__ **Demonstration, Spiraling Winds, TE** Students observe what happens when rising air currents strike a paper spiral. **(GENERAL)**

__ **Discussion, Highs and Lows, TE** Students discuss high and low pressure systems and how they relate to cyclones and anticyclones. **(GENERAL)**

__ **Demonstration, Updrafts, TE** Students observe the lifting power of strong winds. **(GENERAL)**

__ **Math Connection, Coming or Going? TE** Students calculate the difference in speed between lightning and thunder to estimate a thunderstorm's distance. **(GENERAL)**

__ **Teaching Transparency 120, Anatomy of a Hurricane with Worksheet** Use this Teaching Transparency and the accompanying worksheet to support your classroom discussion of the anatomy of a hurricane.

__ **Graphic Organizer, Venn Diagram, SE** Use the graphic organizer to help students develop their analytical skills. **(GENERAL)**

__ **Maps in Action, Weather-Related Disasters, 1980–2003, SE** Students enhance their map-reading skills by answering questions about a map showing the types and locations of weather-related disasters in the United States. **(GENERAL)**

__ **Teaching Transparency 123, Weather-Related Disasters, 1980–2003 with Worksheet** Use this Teaching Transparency and the accompanying worksheet to help students perform the Maps in Action activity.

KEY **SE** = Student Edition **TE** = Teacher Edition **CRF** = Chapter Resource File

__ **Career Focus, Meteorologist, SE** A meteorologist discusses her work studying and researching hurricanes in order to improve forecasting.

__ **Activity, Hurricane Hunters, TE** Students investigate different jobs that support the National Hurricane Center. **(GENERAL)**

__ **Interactive Tutor CD-ROM, Tornadoes and Hurricanes** Students practice what they have learned using this interactive activity. **(GENERAL)**

__ **Interactive Tutor CD-ROM, Lightning** Students practice what they have learned using this interactive activity. **(GENERAL)**

__ **Interactive Tutor CD-ROM, Other Storms and El Niño** Students practice what they have learned using this interactive activity. **(GENERAL)**

CLOSE *(10 minutes)*

__ **Section Review, SE** Students answer end-of-section vocabulary, key ideas, critical thinking, and concept mapping questions. **(GENERAL)**

__ **Reteaching, Game Show, TE** In groups, students write clues that describe the weather patterns discussed in this section and try to guess the patterns described. **(BASIC)**

__ **Quiz, TE** Students answer 3 questions about this section. **(GENERAL)**

__ **Section Quiz, CRF** Use this quiz to assess students' understanding of the section. **(GENERAL) Also in Spanish**

__ **Spanish Section Quiz, Spanish Assessments** Use this quiz to assess students' understanding of the section. **(GENERAL)**

__ **Alternative Assessment, Radio Play, TE** Students develop a radio script that describes the development of a complex weather system. **(GENERAL)**

KEY **SE** = Student Edition **TE** = Teacher Edition **CRF** = Chapter Resource File

Section 3 Weather Instruments

PACING

Regular Schedule: **with lab(s): N/A** **without lab(s): 2 days**
Block Schedule: **with lab(s): N/A** **without lab(s): 1 day**

OBJECTIVES

1. **Identify** four instruments that measure lower-atmospheric weather conditions.
2. **Describe** how scientists measure conditions in the upper atmosphere.
3. **Explain** how computers help scientists understand weather.

NATIONAL SCIENCE EDUCATION STANDARDS

HNS 2a: Science distinguishes itself from other ways of knowing and from other bodies of knowledge through the use of empirical standards, logical arguments, and skepticism, as scientists strive for the best possible explanations about the natural world.

SAI 2a: Scientists usually inquire about how physical, living, or designed systems function. Conceptual principles and knowledge guide scientific inquiries. Historical and current scientific knowledge influence the design and interpretation of investigations and the evaluation of proposed explanations made by other scientists.

UCP 1: The natural and designed world is complex; it is too large and complicated to investigate and comprehend all at once. Scientists and students learn to define small portions for the convenience of investigation. The units of investigation can be referred to as "systems." A system is an organized group of related objects or components that form a whole. Systems can consist, for example, of organisms, machines, fundamental particles, galaxies, ideas, numbers, transportation, and education. Systems have boundaries, components, resources flow (input and output), and feedback.

UCP 2: Evidence consists of observations and data on which to base scientific explanations. Using evidence to understand interactions allows individuals to predict changes in natural and designed systems. Models are tentative schemes or structures that correspond to real objects, events, or classes of events, and that have explanatory power. Models help scientists and engineers understand how things work. Models take many forms, including physical objects, plans, mental constructs, mathematical equations, and computer simulations.

UCP 3: Although most things are in the process of becoming different—changing—some properties of objects and processes are characterized by constancy, including the speed of light, the charge of an electron, and the total mass plus energy in the universe. Changes might occur, for example, in properties of materials, position of objects, motion, and form and function of systems. Interactions within and among systems result in change. Changes vary in rate, scale, and pattern, including trends and cycles.

KEY **SE** = Student Edition **TE** = Teacher Edition **CRF** = Chapter Resource File

Lesson Plan

FOCUS *(5 minutes)*

__ **Overview, TE** Review the objectives listed in the Student Edition. **(GENERAL)**

__ **PowerPoint® Resources** Use the customizable presentation to enhance your lecture.

__ **Bellringer, TE** Students brainstorm a list of weather conditions that scientists monitor. **(GENERAL)**

__ **Bellringer Transparency** Use this transparency as students enter the classroom and find their seats.

MOTIVATE *(10 minutes)*

__ **Group Activity, Simple Thermometer, TE** Students observe how heat causes liquid to expand and rise. **(GENERAL)**

TEACH *(20 minutes)*

__ **Directed Reading, CRF** These worksheets reinforce basic concepts and vocabulary presented in the section. **(BASIC)**

__ **Quick Lab, Wind Chill, SE** Students observe how moving air affects the temperature of water. **(GENERAL)**

__ **Datasheet for Quick Lab, Wind Chill, CRF** Students use the datasheet to complete the Quick Lab. **(GENERAL)**

__ **Skill Builder, Graphing, TE** Students take hourly temperature readings and graph the results to identify temperature patterns. **(GENERAL)**

__ **Internet Activity, Weather Images, TE** Students visit the National Weather Service Web site and research a topic that interests them. **(GENERAL)**

__ **Internet Activity, Weather Images, CRF** This worksheet directs student research for the topic presented in the Teacher Edition. **(GENERAL)**

__ **Mapping Expeditions, Snapshots of the Weather** Students study a series of weather maps and make predictions based on their observations. **(GENERAL)**

__ **Inquiry Lab, Building a Weather Station, CRF** Students build a weather station to practice using cloud formations and barometric pressure to track and predict weather changes. **(GENERAL)**

CLOSE *(10 minutes)*

__ **Section Review, SE** Students answer end-of-section vocabulary, key ideas, critical thinking, and concept mapping questions. **(GENERAL)**

__ **Reteaching, Tools of the Trade, TE** Students create posters illustrating instruments used to measure weather variables and how those tools work. **(BASIC)**

__ **Quiz, TE** Students answer 3 questions about this section. **(GENERAL)**

__ **Section Quiz, CRF** Use this quiz to assess students' understanding of the section. **(GENERAL) Also in Spanish**

KEY **SE** = Student Edition **TE** = Teacher Edition **CRF** = Chapter Resource File

__ **Spanish Section Quiz, Spanish Assessments** Use this quiz to assess students' understanding of the section. **(GENERAL)**

__ **Alternative Assessment, Weather Riddles, TE** Students create booklets containing riddles that they have written to describe a weather instrument. **(GENERAL)**

KEY **SE** = Student Edition **TE** = Teacher Edition **CRF** = Chapter Resource File

Section 4 Forecasting the Weather

PACING

Regular Schedule:	**with lab(s): 2 days**	**without lab(s): 1 day**
Block Schedule:	**with lab(s): 1 day**	**without lab(s): 0.5 day**

OBJECTIVES

1. **Explain** how weather stations communicate weather data.
2. **Explain** how a weather map is created.
3. **Explain** how computer models help meteorologists forecast weather.
4. **List** three types of weather that meteorologists have attempted to control.

NATIONAL SCIENCE EDUCATION STANDARDS

HNS 2a: Science distinguishes itself from other ways of knowing and from other bodies of knowledge through the use of empirical standards, logical arguments, and skepticism, as scientists strive for the best possible explanations about the natural world.

SAI 2a: Scientists usually inquire about how physical, living, or designed systems function. Conceptual principles and knowledge guide scientific inquiries. Historical and current scientific knowledge influence the design and interpretation of investigations and the evaluation of proposed explanations made by other scientists.

SPSP 1a: Hazards and the potential for accidents exist. Regardless of the environment, the possibility of injury, illness, disability, or death may be present. Humans have a variety of mechanisms—sensory, motor, emotional, social, and technological—that can reduce and modify hazard

UCP 1: The natural and designed world is complex; it is too large and complicated to investigate and comprehend all at once. Scientists and students learn to define small portions for the convenience of investigation. The units of investigation can be referred to as "systems." A system is an organized group of related objects or components that form a whole. Systems can consist, for example, of organisms, machines, fundamental particles, galaxies, ideas, numbers, transportation, and education. Systems have boundaries, components, resources flow (input and output), and feedback.

UCP 2: Evidence consists of observations and data on which to base scientific explanations. Using evidence to understand interactions allows individuals to predict changes in natural and designed systems. Models are tentative schemes or structures that correspond to real objects, events, or classes of events, and that have explanatory power. Models help scientists and engineers understand how things work. Models take many forms, including physical objects, plans, mental constructs, mathematical equations, and computer simulations.

UCP 3: Although most things are in the process of becoming different—changing—some properties of objects and processes are characterized by constancy, including the

KEY **SE** = Student Edition **TE** = Teacher Edition **CRF** = Chapter Resource File

speed of light, the charge of an electron, and the total mass plus energy in the universe. Changes might occur, for example, in properties of materials, position of objects, motion, and form and function of systems. Interactions within and among systems result in change. Changes vary in rate, scale, and pattern, including trends and cycles.

FOCUS *(5 minutes)*

__ **Overview, TE** Review the objectives listed in the Student Edition. **(GENERAL)**

__ **PowerPoint® Resources** Use the customizable presentation to enhance your lecture.

__ **Bellringer, TE** Students write down the kind of information they would expect a weather forecast to include. **(GENERAL)**

__ **Bellringer Transparency** Use this transparency as students enter the classroom and find their seats.

MOTIVATE *(10 minutes)*

__ **Discussion, Weather Lore, TE** Students discuss stories and sayings from other cultures about the weather. **(GENERAL)**

TEACH *(65 minutes)*

__ **Directed Reading, CRF** These worksheets reinforce basic concepts and vocabulary presented in the section. **(BASIC)**

__ **Teaching Transparency 121, Weather Symbols with Worksheet** Use this Teaching Transparency and the accompanying worksheet to support your classroom discussion of weather symbols.

__ **Using the Figure, Classroom Station Model, TE** Students create a classroom station model as part of a weather bulletin board. **(GENERAL)**

__ **Discussion, Comprehension Check, TE** Students discuss how to determine the likelihood of precipitation. **(BASIC)**

__ **Group Activity, Cloud Cover, TE** Students use mirrors to determine cloud cover percentage and upper atmospheric wind direction. **(GENERAL)**

__ **Discussion, Barometric Pressure, TE** Students discuss what rising and falling barometric pressure indicates. **(GENERAL)**

__ **Activity, Local Weather Maps, TE** Students bring in weather maps and list the symbols they find. **(GENERAL)**

__ **Using the Figure, Isobars, TE** Students discuss how low- and high-pressure data helps forecast the weather. **(GENERAL)**

__ **Discussion, Isotherms, TE** Students discuss which weather characteristic maps with isotherms help to locate. **(BASIC)**

__ **Teaching Transparency 122, Weather Map of the United States with Worksheet** Use this Teaching Transparency and the accompanying worksheet to support your classroom discussion of weather maps of the United States.

KEY **SE** = Student Edition **TE** = Teacher Edition **CRF** = Chapter Resource File

__ **Quick Lab, Gathering Weather Data, SE** Students determine the weather conditions at the school and create a station model to describe the findings. **(GENERAL)**

__ **Datasheet for Quick Lab, Gathering Weather Data, CRF** Students use the datasheet to complete the Quick Lab. **(GENERAL)**

__ **Literature Connection, The Perfect Storm, TE** Students read *The Perfect Storm* and prepare a report on the material. **(GENERAL)**

__ **Group Activity, Weather Safety Project, TE** Students explore ways of staying safe during various weather emergencies. **(GENERAL)**

__ **Debate, Weather Modification, TE** Students research and debate ethical and environmental issues surrounding weather modification experiments. **(ADVANCED)**

__ **Skills Practice Lab, Weather Map Interpretation, SE** Students study weather map symbols to understand the relationships between temperature, pressure, and winds. **(GENERAL)**

__ **Datasheet for Chapter Lab, Skills Practice Lab, Weather Map Interpretation, CRF** Students use the datasheet to complete the Chapter Lab. **(GENERAL)**

__ **Making Models Lab, Blowing in the Wind, CRF** Students examine how temperature differences cause winds and build a wind gauge to study wind strength and direction. **(GENERAL)**

__ **Interactive Tutor CD-ROM, Old-Time Forecasting** Students practice what they have learned using this interactive activity. **(GENERAL)**

__ **Interactive Tutor CD-ROM, Modern Forecasting** Students practice what they have learned using this interactive activity. **(GENERAL)**

CLOSE *(10 minutes)*

__ **Section Review, SE** Students answer end-of-section vocabulary, key ideas, critical thinking, and concept mapping questions. **(GENERAL)**

__ **Homework, Weather Station, TE** Students build a class weather station and use it to collect data and make weather predictions. **(ADVANCED)**

__ **Reteaching, Matching Game, TE** Students draw meteorological symbols on index cards and use them to play a matching game. **(BASIC)**

__ **Quiz, TE** Students answer 5 questions about this section. **(GENERAL)**

__ **Section Quiz, CRF** Use this quiz to assess students' understanding of the section. **(GENERAL) Also in Spanish**

__ **Spanish Section Quiz, Spanish Assessments** Use this quiz to assess students' understanding of the section. **(GENERAL)**

__ **Alternative Assessment, Station Models, TE** Students create a station model to record data from a local weather report. **(GENERAL)**

KEY **SE** = Student Edition **TE** = Teacher Edition **CRF** = Chapter Resource File

End of Chapter Review and Assessment

PACING

Regular Schedule:	**with lab(s): N/A**	**without lab(s): 2 days**
Block Schedule:	**with lab(s): N/A**	**without lab(s): 1 day**

__ **Chapter Review, SE** Students answer end-of-chapter vocabulary, key ideas, critical thinking, math, writing, and interpreting graphics questions. **(GENERAL)**

__ **Alternative Assessment, Weatherwise Museum, TE** In groups, students create models, diagrams, and activities to explain the concepts of this chapter. **(GENERAL)**

__ **Standardized Test Prep, SE** Have students complete the standardized test prep to help them prepare for standardized tests. **(GENERAL)**

__ **Holt PuzzlePro®** Use this software to create crossword puzzles and word searches that make learning vocabulary fun. **(GENERAL)**

__ **Concept Review, CRF** Use this worksheet to assess student's understanding of the main concepts presented in the chapter. **(GENERAL)**

__ **Critical Thinking, CRF** Students use this worksheet to develop and practice their critical thinking skills. **(ADVANCED)**

__ **Math Skills, CRF** This worksheet helps reinforce students' understanding of order of operations. **(GENERAL)**

__ **Graphing Skills, CRF** This worksheet helps reinforce students' understanding of bar graphs. **(GENERAL)**

__ **Concept Review, Study Guide, Weather** Use this worksheet to assess student's understanding of the main concepts presented in the chapter. **(GENERAL)**

__ **Concept Review, Spanish Study Guide, Weather** This worksheet helps reinforce English-language learners' understanding of the main concepts presented in the chapter. **(GENERAL)**

__ **Chapter Test A, CRF** Assign questions from the appropriate test for chapter assessment. **(GENERAL)**

__ **Chapter Test B, CRF** Assign questions from the appropriate test for chapter assessment. **(ADVANCED)**

__ **Spanish Chapter Test A, Spanish Assessments** Assign questions from the appropriate test for chapter assessment. **(GENERAL)**

__ **Test Generator, One-Stop Planner** Create a customized homework assignment, quiz, or test using the HRW Test Generator program. **(GENERAL)**

__ **Test Item Listing, CRF** Use the Test Item Listing to identify questions for a customized homework, quiz, or test.

__ **PowerPoint® Resources** Use the customizable presentation to help students practice standardized test-taking skills.

KEY **SE** = Student Edition **TE** = Teacher Edition **CRF** = Chapter Resource File

OTHER RESOURCE OPTIONS

__ **NOVA Video, "Hurricane!"** Watch scientists fly into the centers of hurricanes so that they can learn more about the hurricanes, and listen to firsthand accounts of people affected by Hurricane Camille.

__ **Brain Food Video Quiz** Use this video resource to assess student's progress. **(GENERAL)**

KEY **SE** = Student Edition **TE** = Teacher Edition **CRF** = Chapter Resource File

Chapter 25 Climate

Chapter Opener

__ **Pre-Reading Activity, Tri-Fold, SE**

__ **Using the Figure, Life Adapts, TE (GENERAL)**

__ **Long-Term Project, Comparing Climate Features, SE** Students keep a daily weather log for six months in order to draw conclusions about their climate and compare it to other regions. **(ADVANCED)**

__ **Long-Term Project, Comparing Climate Features, Long-term Projects** Students use this datasheet to complete the Long-Term Project. **(ADVANCED)**

Section 1 Factors That Affect Climate

PACING

Regular Schedule:	**with lab(s): 2 days**	**without lab(s): 1 day**
Block Schedule:	**with lab(s): 1 day**	**without lab(s): 0.5 day**

OBJECTIVES

1. **Identify** two major factors used to describe climate.
2. **Explain** how latitude determines the amount of solar energy received on Earth.
3. **Describe** how the different rates at which land and water are heated affect climate.
4. **Explain** the effects of topography on climate.

NATIONAL SCIENCE EDUCATION STANDARDS

ES 1d: Global climate is determined by energy transfer from the sun at and near the earth's surface. This energy transfer is influenced by dynamic processes such as cloud cover and the earth's rotation, and static conditions such as the position of mountain ranges and oceans.

FOCUS *(5 minutes)*

__ **Overview, TE** Review the objectives listed in the Student Edition. **(GENERAL)**

__ **PowerPoint® Resources** Use the customizable presentation to enhance your lecture.

__ **Bellringer, TE** Students discuss the difference between weather and climate. **(GENERAL)**

__ **Bellringer Transparency** Use this transparency as students enter the classroom and find their seats.

MOTIVATE *(10 minutes)*

__ **Using the Figure, Temperature Ranges, TE** Students determine the average yearly temperatures and yearly temperature ranges for St. Louis and San Francisco. **(GENERAL)**

KEY **SE** = Student Edition **TE** = Teacher Edition **CRF** = Chapter Resource File

Lesson Plan

TEACH *(65 minutes)*

__ **Directed Reading, CRF** These worksheets reinforce basic concepts and vocabulary presented in the section. **(BASIC)**

__ **Teaching Transparency 124, Average Sea-Level Temperatures During Winter in the Northern Hemisphere with Worksheet** Use this Teaching Transparency and the accompanying worksheet to support your classroom discussion of average sea-level temperatures during winter in the northern hemisphere.

__ **Reading Skill Builder, Paired Summarizing, TE** Students read the text in pairs and take turns summarizing what they have read. **(BASIC)**

__ **Demonstration, Latitude and Temperature, TE** Students study how latitude affects temperature by placing thermometers at various angles to a lamp. **(GENERAL)**

__ **Using the Figure, Shifting Winds, TE** Students discuss why climates near the equator do not have temperature-based seasons. **(ADVANCED)**

__ **Quick Lab, Evaporation, SE** Students use lamps to determine the relationship between temperature and the rate of evaporation. **(GENERAL)**

__ **Datasheet for Quick Lab, Evaporation, CRF** Students use the datasheet to complete the Quick Lab. **(GENERAL)**

__ **Activity, El Niño, TE** Students research El Niño events, how they are predicted, and their effects on wildlife and weather patterns. **(GENERAL)**

__ **Inquiry Lab, Factors That Affect Climate, SE** Students explore how properties of land and water affect climate. **(GENERAL)**

__ **Datasheet for Chapter Lab, Inquiry Lab, Factors That Affect Climate, CRF** Students use the datasheet to complete the Chapter Lab. **(GENERAL)**

__ **Maps in Action, Climates of the World, SE** Students enhance their map-reading skills by answering questions about a map showing the climates of the world.

__ **Teaching Transparency 129, Climates of the World with Worksheet** Use this Teaching Transparency and the accompanying worksheet to help students perform the Maps in Action activity.

__ **Discussion, Climate Classification, TE** Students discuss classification schemes and research other climate classification systems. **(GENERAL)**

__ **Mapping Expeditions, Where the Hippos Roam, SE** Students use data from a fossil site, create a map of that site, and draw conclusions about its past environment. **(GENERAL)**

__ **Interactive Tutor CD-ROM, Other Storms and El Niño** Students practice what they have learned using this interactive activity. **(GENERAL)**

__ **Interactive Tutor CD-ROM, Greenhouse Effect** Students practice what they have learned using this interactive activity. **(GENERAL)**

CLOSE *(10 minutes)*

__ **Section Review, SE** Students answer end-of-section vocabulary, key ideas, critical thinking, and concept mapping questions. **(GENERAL)**

KEY **SE** = Student Edition **TE** = Teacher Edition **CRF** = Chapter Resource File

__ **Reteaching, Climate Factors, TE** Students write descriptions of factors that affect climate on index cards and quiz each other. **(BASIC)**

__ **Quiz, TE** Students answer 2 questions about this section. **(GENERAL)**

__ **Section Quiz, CRF** Use this quiz to assess students' understanding of the section. **(GENERAL) Also in Spanish**

__ **Spanish Section Quiz, Spanish Assessments** Use this quiz to assess students' understanding of the section. **(GENERAL)**

__ **Alternative Assessment, Graphing, TE** Students graph the monthly average precipitation and average high and low temperatures for their city. **(GENERAL)**

KEY **SE** = Student Edition **TE** = Teacher Edition **CRF** = Chapter Resource File

Section 2 Climate Zones

PACING

Regular Schedule:	**with lab(s): N/A**	**without lab(s): 2 days**
Block Schedule:	**with lab(s): N/A**	**without lab(s): 1 day**

OBJECTIVES

1. **Describe** the three types of tropical climates.
2. **Describe** the five types of middle-latitude climates.
3. **Describe** the three types of polar climates
4. **Explain** why city climates may differ from rural climates.

NATIONAL SCIENCE EDUCATION STANDARDS

ES 1d: Global climate is determined by energy transfer from the sun at and near the earth's surface. This energy transfer is influenced by dynamic processes such as cloud cover and the earth's rotation, and static conditions such as the position of mountain ranges and oceans.

SPSP 5b: Human activities can enhance potential for hazards. Acquisition of resources, urban growth, and waste disposal can accelerate rates of natural change.

FOCUS *(5 minutes)*

__ **Overview, TE** Review the objectives listed in the Student Edition. **(GENERAL)**

__ **PowerPoint® Resources** Use the customizable presentation to enhance your lecture.

__ **Bellringer, TE** Students describe the climate and physical features of their area and list other regions that have similar climates. **(GENERAL)**

__ **Bellringer Transparency** Use this transparency as students enter the classroom and find their seats.

MOTIVATE *(10 minutes)*

__ **Discussion, Name That Climate, TE** Students list and describe as many climates as they can think of. **(GENERAL)**

TEACH *(20 minutes)*

__ **Directed Reading, CRF** These worksheets reinforce basic concepts and vocabulary presented in the section. **(BASIC)**

__ **Teaching Transparency 125, Tropical Climates with Worksheet** Use this Teaching Transparency and the accompanying worksheet to support your classroom discussion of tropical climates.

KEY **SE** = Student Edition **TE** = Teacher Edition **CRF** = Chapter Resource File

__ **Activity, Adaptations, TE** Students study photos of plants, determine where the plants are most likely to grow, and discuss adaptations that enable them to flourish. **(GENERAL)**

__ **Skill Builder, Math, TE** Students convert temperature and precipitation values from SI units to English units. **(GENERAL)**

__ **Teaching Transparency 126, Middle-Latitude Climates with Worksheet** Use this Teaching Transparency and the accompanying worksheet to support your classroom discussion of middle-latitude climates.

__ **Teaching Transparency 127, Polar Climates with Worksheet** Use this Teaching Transparency and the accompanying worksheet to support your classroom discussion of polar climates.

__ **Skills Practice Lab, Microclimates, CRF** Students study different areas to determine their microclimate and relate their findings to the organisms living there. **(GENERAL)**

__ **Interactive Tutor CD-ROM, Climate Zones** Students practice what they have learned using this interactive activity. **(GENERAL)**

CLOSE *(10 minutes)*

__ **Section Review, SE** Students answer end-of-section vocabulary, key ideas, critical thinking, and concept mapping questions. **(GENERAL)**

__ **Reteaching, Flipbook, TE** Students write all of the subclimates on index cards and organize them into major climate zones. **(BASIC)**

__ **Quiz, TE** Students answer 2 questions about this section. **(GENERAL)**

__ **Section Quiz, CRF** Use this quiz to assess students' understanding of the section. **(GENERAL) Also in Spanish**

__ **Spanish Section Quiz, Spanish Assessments** Use this quiz to assess students' understanding of the section. **(GENERAL)**

__ **Alternative Assessment, Travel Agent, TE** Students choose a destination and create a brochure describing the destination. **(GENERAL)**

KEY **SE** = Student Edition **TE** = Teacher Edition **CRF** = Chapter Resource File

Section 3 Climate Change

PACING

Regular Schedule:	**with lab(s): N/A**	**without lab(s): 2 days**
Block Schedule:	**with lab(s): N/A**	**without lab(s): 1 day**

OBJECTIVES

1. **Compare** four methods used to study climate change.
2. **Describe** four factors that may cause climate change.
3. **Identify** potential impacts of climate change.
4. **Identify** ways that humans can minimize their effect on climate change.

NATIONAL SCIENCE EDUCATION STANDARDS

ES 1d: Global climate is determined by energy transfer from the sun at and near the earth's surface. This energy transfer is influenced by dynamic processes such as cloud cover and the earth's rotation, and static conditions such as the position of mountain ranges and oceans.

SPSP 5b: Human activities can enhance potential for hazards. Acquisition of resources, urban growth, and waste disposal can accelerate rates of natural change.

SPSP 4a: Natural ecosystems provide an array of basic processes that affect humans. Those processes include maintenance of the quality of the atmosphere, generation of soils, control of the hydrologic cycle, disposal of wastes, and recycling of nutrients. Humans are changing many of these basic processes, and the changes may be detrimental to humans.

SPSP 4b: Materials from human societies affect both physical and chemical cycles of the earth.

SPSP 4c: Many factors influence environmental quality. Factors that students might investigate include population growth, resource use, population distribution, overconsumption, the capacity of technology to solve problems, poverty, the role of economic, political, and religious views, and different ways humans view the earth.

SPSP 5a: Normal adjustments of earth may be hazardous for humans. Humans live at the interface between the atmosphere driven by solar energy and the upper mantle where convection creates changes in the earth's solid crust. As societies have grown, become stable, and come to value aspects of the environment, vulnerability to natural processes of change has increased.

SPSP 5d: Natural and human-induced hazards present the need for humans to assess potential danger and risk. Many changes in the environment designed by humans bring benefits to society, as well as cause risks. Students should understand the costs and trade-offs of various hazards—ranging from those with minor risk to a few people to major catastrophes with major risk to many people. The scale of events and the accuracy with

KEY **SE** = Student Edition **TE** = Teacher Edition **CRF** = Chapter Resource File

which scientists and engineers can (and cannot) predict events are important considerations.

UCP 2: Evidence consists of observations and data on which to base scientific explanations. Using evidence to understand interactions allows individuals to predict changes in natural and designed systems. Models are tentative schemes or structures that correspond to real objects, events, or classes of events, and that have explanatory power. Models help scientists and engineers understand how things work. Models take many forms, including physical objects, plans, mental constructs, mathematical equations, and computer simulations.

FOCUS *(5 minutes)*

__ **Overview, TE** Review the objectives listed in the Student Edition. **(GENERAL)**

__ **PowerPoint® Resources** Use the customizable presentation to enhance your lecture.

__ **Bellringer, TE** Students write down five ways that their region would be affected by cooler temperatures. **(GENERAL)**

__ **Bellringer Transparency** Use this transparency as students enter the classroom and find their seats.

MOTIVATE *(10 minutes)*

__ **Activity, Tree Rings, TE** Students examine a cross–section of a tree trunk and discuss conditions that would produce thicker or thinner rings. **(GENERAL)**

TEACH *(20 minutes)*

__ **Directed Reading, CRF** These worksheets reinforce basic concepts and vocabulary presented in the section. **(BASIC)**

__ **Internet Activity, Climate Models, TE** Students visit government Web sites and research general circulation models. **(GENERAL)**

__ **Internet Activity, Climate Models, CRF** This worksheet directs student research for the topic presented in the Teacher Edition. **(GENERAL)**

__ **Debate, Global Warming, TE** Students research and debate human influences on global warming. **(ADVANCED)**

__ **Graphic Organizer, Cause–and–Effect Map, SE** Use the graphic organizer to help students develop their analytical skills. **(GENERAL)**

__ **Teaching Transparency 128, Orbital Changes and Climate with Worksheet** Use this Teaching Transparency and the accompanying worksheet to support your classroom discussion of orbital changes and climate.

__ **Quick Lab, Hot Stuff, SE** Students compare air temperatures over a grassy area and a parking lot. **(GENERAL)**

__ **Datasheet for Quick Lab, Hot Stuff, CRF** Students use the datasheet to complete the Quick Lab. **(GENERAL)**

KEY **SE** = Student Edition **TE** = Teacher Edition **CRF** = Chapter Resource File

__ **Demonstration, Icecaps Melting, TE** Students observe how melting icecaps affect water levels. **(BASIC)**

__ **Law Connection, Kyoto Protocol, TE** Students research the status of the Kyoto Protocol and the position of the United States regarding the treaty. **(ADVANCED)**

__ **Using the Figure, Rising Seas, TE** Students research the causes and consequences of rising sea levels. **(GENERAL)**

__ **Reading Skill Builder, Reading Organizer, TE** Students are encouraged to take notes while reading this section. **(BASIC)**

__ **Eye on the Environment, Keeping Cool with Algae, SE** Coccolithophores, algae microorganisms, absorb carbon dioxide and effectively impact the Greenhouse Effect.

__ **Internet Activity, Gaia Hypothesis, TE** Students research the Gaia hypothesis and discuss whether it is helpful in understanding how climates work. **(GENERAL)**

__ **Internet Activity, Gaia Hypothesis, CRF** This worksheet directs student research for the topic presented in the Teacher Edition. **(GENERAL)**

__ **Inquiry Lab, Particulates in the Atmosphere, CRF** Students study how particulates in the atmosphere affect temperature and climate. **(GENERAL)**

__ **Interactive Tutor CD-ROM, Past Climate** Students practice what they have learned using this interactive activity. **(GENERAL)**

__ **Interactive Tutor CD-ROM, Recent Climate** Students practice what they have learned using this interactive activity. **(GENERAL)**

CLOSE *(10 minutes)*

__ **Section Review, SE** Students answer end-of-section vocabulary, key ideas, critical thinking, and concept mapping questions. **(GENERAL)**

__ **Homework, Volcanoes and Climate, TE** Students research major historical volcanic eruptions and create a timeline documenting their effects on climate. **(ADVANCED)**

__ **Reteaching, Peer Reviewing, TE** Students compose questions about the material in this section and quiz each other. **(BASIC)**

__ **Quiz, TE** Students answer 2 questions about this section. **(GENERAL)**

__ **Section Quiz, CRF** Use this quiz to assess students' understanding of the section. **(GENERAL) Also in Spanish**

__ **Spanish Section Quiz, Spanish Assessments** Use this quiz to assess students' understanding of the section. **(GENERAL)**

__ **Alternative Assessment, Global Warming in the News, TE** Students write newspaper articles about how scientists study and monitor global warming. **(GENERAL)**

KEY **SE** = Student Edition **TE** = Teacher Edition **CRF** = Chapter Resource File

End of Chapter Review and Assessment

PACING

Regular Schedule:	**with lab(s): N/A**	**without lab(s): 2 days**
Block Schedule:	**with lab(s): N/A**	**without lab(s): 1 day**

__ **Chapter Review, SE** Students answer end-of-chapter vocabulary, key ideas, critical thinking, math, writing, and interpreting graphics questions. **(GENERAL)**

__ **Alternative Assessment, Subclimate Story, TE** Students choose a subclimate other than their own and write a story about a family living there. **(ADVANCED)**

__ **Standardized Test Prep, SE** Have students complete the standardized test prep to help them prepare for standardized tests. **(GENERAL)**

__ **Holt PuzzlePro®** Use this software to create crossword puzzles and word searches that make learning vocabulary fun. **(GENERAL)**

__ **Concept Review, CRF** Use this worksheet to assess student's understanding of the main concepts presented in the chapter. **(GENERAL)**

__ **Critical Thinking, CRF** Students use this worksheet to develop and practice critical thinking skills. **(ADVANCED)**

__ **Math Skills, CRF** This worksheet helps reinforce students' understanding of geometry. **(GENERAL)**

__ **Graphing Skills, CRF** This worksheet helps reinforce students' understanding of bar graphs. **(GENERAL)**

__ **Concept Review, Study Guide, Climate** Use this worksheet to assess student's understanding of the main concepts presented in the chapter. **(GENERAL)**

__ **Concept Review, Spanish Study Guide, Climate** This worksheet helps reinforce English-language learners' understanding of the main concepts presented in the chapter. **(GENERAL)**

__ **Chapter Test A, CRF** Assign questions from the appropriate test for chapter assessment. **(GENERAL)**

__ **Chapter Test B, CRF** Assign questions from the appropriate test for chapter assessment. **(ADVANCED)**

__ **Spanish Chapter Test A, Spanish Assessments** Assign questions from the appropriate test for chapter assessment. **(GENERAL)**

__ **Test Generator, One-Stop Planner** Create a customized homework assignment, quiz, or test using the HRW Test Generator program. **(GENERAL)**

__ **Test Item Listing, CRF** Use the Test Item Listing to identify questions for a customized homework, quiz, or test.

__ **PowerPoint® Resources** Use the customizable presentation to help students practice standardized test-taking skills.

KEY **SE** = Student Edition **TE** = Teacher Edition **CRF** = Chapter Resource File

OTHER RESOURCE OPTIONS

__ **Brain Food Video Quiz** Use this video resource to assess student's progress. **(GENERAL)**

KEY **SE** = Student Edition **TE** = Teacher Edition **CRF** = Chapter Resource File

Chapter 26 Studying Space

Chapter Opener

__ **Pre-Reading Activity, Two-Panel Flip Chart, SE**

__ **Using the Figure, Man in Space, TE (GENERAL)**

__ **Long-Term Project, Apparent Motions of the Moon, Long-Term Projects** Students make daily observations of the moon to determine the path and phases of the moon. **(ADVANCED)**

__ **Long-Term Project, Positions of Sunrise and Sunset, SE** Students measure changes in the direction of sunrise and sunset for 8 or 9 months. **(ADVANCED)**

__ **Long-Term Project, Positions of Sunrise and Sunset, CRF** Students use this datasheet to complete the Long-Term Project. **(ADVANCED)**

Section 1 Viewing the Universe

PACING

Regular Schedule:	**with lab(s): N/A**	**without lab(s): 2 days**
Block Schedule:	**with lab(s): N/A**	**without lab(s): 1 day**

OBJECTIVES

1. **Describe** characteristics of the universe in terms of time, distance, and organization.
2. **Identify** the visible and nonvisible parts of the electromagnetic spectrum.
3. **Compare** refracting telescopes and reflecting telescopes.
4. **Explain** how telescopes for nonvisible electromagnetic radiation differ from light telescopes.

NATIONAL SCIENCE EDUCATION STANDARDS

ES 4a: The origin of the universe remains one of the greatest questions in science. The "big bang" theory places the origin between 10 and 20 billion years ago, when the universe began in a hot dense state; according to this theory, the universe has been expanding ever since.

ES 4b: Early in the history of the universe, matter, primarily the light atoms hydrogen and helium, clumped together by gravitational attraction to form countless trillions of stars. Billions of galaxies, each of which is a gravitationally bound cluster of billions of stars, now form most of the visible mass in the universe.

HNS 3c: Occasionally, there are advances in science and technology that have important and long-lasting effects on science and society. Examples of such advances include the following: Copernican revolution; Newtonian mechanics; Relativity; Geologic time scale; Plate tectonics; Atomic theory; Nuclear physics; Biological evolution; Germ theory;

KEY **SE** = Student Edition **TE** = Teacher Edition **CRF** = Chapter Resource File

Industrial revolution; Molecular biology; Information and communication; Quantum theory; Galactic universe; Medical and health technology.

PS 5a: The total energy of the universe is constant. Energy can be transferred by collisions in chemical and nuclear reactions, by light waves and other radiations, and in many other ways. However, it can never be destroyed. As these transfers occur, the matter involved becomes steadily less ordered.

SAI 2c: Scientists rely on technology to enhance the gathering and manipulation of data. New techniques and tools provide new evidence to guide inquiry and new methods to gather data, thereby contributing to the advance of science. The accuracy and precision of the data, and therefore the quality of the exploration, depends on the technology used.

SAI 2d: Mathematics is essential in scientific inquiry. Mathematical tools and models guide and improve the posing of questions, gathering data, constructing explanations and communicating results.

SPSP 6b: Understanding basic concepts and principles of science and technology should precede active debate about the economics, policies, politics, and ethics of various science- and technology-related challenges. However, understanding science alone will not resolve local, national, or global challenges.

SPSP 6c: Progress in science and technology can be affected by social issues and challenges. Funding priorities for specific health problems serve as examples of ways that social issues influence science and technology.

SPSP 6d: Individuals and society must decide on proposals involving new research and the introduction of new technologies into society. Decisions involve assessment of alternatives, risks, costs, and benefits and consideration of who benefits and who suffers, who pays and gains, and what the risks are and who bears them. Students should understand the appropriateness and value of basic questions—"What can happen?"—"What are the odds?"—and "How do scientists and engineers know what will happen?"

UCP: Unifying Concepts and Processes

FOCUS *(5 minutes)*

__ **Overview, TE** Review the objectives listed in the Student Edition. **(GENERAL)**

__ **PowerPoint® Resources** Use the customizable presentation to enhance your lecture.

__ **Bellringer, TE** Students draw pictures of astronomers at work. **(GENERAL)**

__ **Bellringer Transparency** Use this transparency as students enter the classroom and find their seats.

MOTIVATE *(10 minutes)*

__ **Activity, Descriptive Writing, TE** Students write essays comparing astronomy with ancient space science. **(GENERAL)**

KEY **SE** = Student Edition **TE** = Teacher Edition **CRF** = Chapter Resource File

TEACH *(20 minutes)*

__ **Directed Reading, CRF** These worksheets reinforce basic concepts and vocabulary presented in the section. **(BASIC)**

__ **Group Activity, Cosmic Timeline, TE** Students place major historical events on a timeline to visualize the time span since the beginning of the universe. **(GENERAL)**

__ **Discussion, Galactic Address, TE** Students describe their galactic addresses. **(BASIC)**

__ **Skill Builder, Math, TE** Students calculate how far light travels in one year, then they calculate the distance to Sirius in light-years. **(GENERAL)**

__ **Reading Skill Builder, Paired Summarizing, TE** Students read the text in pairs and take turns summarizing what they have read. **(BASIC)**

__ **Demonstration, Colors of Light, TE** Students observe a spectrum and describe what they see. **(BASIC)**

__ **Group Activity, Splitting Light, TE** Students use diffraction grating to observe the light spectrum. **(GENERAL)**

__ **Demonstration, Heating with Invisible Light, TE** Students observe the thermal properties of infrared radiation. **(GENERAL)**

__ **History Connection, Seeing the Light, TE** Students research scientists that contributed to our understanding of electromagnetic radiation. **(ADVANCED)**

__ **Activity, Telescope Optics, TE** Students use two magnifying glasses to show how telescopes work. **(BASIC)**

__ **Using the Figure, Comparing Telescopes, TE** Students compare refracting and reflecting telescopes. **(GENERAL)**

__ **Debate, Space Exploration, TE** Students debate the pros and cons of human space exploration. **(GENERAL)**

__ **Internet Activity, Space Spinoffs, TE** Students identify products of space technology that now have everyday applications. **(GENERAL)**

__ **Internet Activity, Space Spinoffs, CRF** This worksheet directs student research for the topic presented in the Teacher Edition. **(GENERAL)**

__ **Maps in Action, Light Sources, SE** Students enhance their map-reading skills by answering questions about a map showing light sources on Earth. **(GENERAL)**

__ **Teaching Transparency 135, Light Sources with Worksheet** Use this Teaching Transparency and the accompanying worksheet to help students perform the Maps in Action activity.

__ **Group Activity, Community Survey, TE** Students prepare a survey of the usage of outdoor lights in their community. **(GENERAL)**

__ **Inquiry Lab, Comet Meets Jupiter, CRF** Students create model craters to study what happens when comets or meteors strike planets. **(GENERAL)**

__ **Making Models Lab, Telescopes, CRF** Students construct a telescope using two lenses, a straight pin, and aluminum foil. **(GENERAL)**

KEY **SE** = Student Edition **TE** = Teacher Edition **CRF** = Chapter Resource File

__ **Interactive Tutor CD-ROM, Eyes and Mind** Students practice what they have learned using this interactive activity. **(GENERAL)**

__ **Interactive Tutor CD-ROM, Telescopes and Spectroscopes** Students practice what they have learned using this interactive activity. **(GENERAL)**

__ **Interactive Tutor CD-ROM, Satellite Explorations** Students practice what they have learned using this interactive activity. **(GENERAL)**

__ **Interactive Tutor CD-ROM, Light and Electromagnetic Energy** Students practice what they have learned using this interactive activity. **(GENERAL)**

CLOSE *(10 minutes)*

__ **Section Review, SE** Students answer end-of-section vocabulary, key ideas, critical thinking, and concept mapping questions. **(GENERAL)**

__ **Homework, Space Telescopes and Probes, TE** Students research two of the telescopes or probes mentioned in the text. **(ADVANCED)**

__ **Reteaching, Two-Column Notes, TE** In pairs, students create charts that outline the idea of observing the universe. **(BASIC)**

__ **Quiz, TE** Students answer 3 questions about this section. **(GENERAL)**

__ **Section Quiz, CRF** Use this quiz to assess students' understanding of the section. **(GENERAL) Also in Spanish**

__ **Spanish Section Quiz, Spanish Assessments** Use this quiz to assess students' understanding of the section. **(GENERAL)**

__ **Alternative Assessment, Skit, TE** Students perform skits about the history of astronomy. **(ADVANCED)**

KEY **SE** = Student Edition **TE** = Teacher Edition **CRF** = Chapter Resource File

Lesson Plan

Section 2 Movements of Earth

PACING

Regular Schedule:	**with lab(s): 2 days**	**without lab(s): 1 day**
Block Schedule:	**with lab(s): 1 day**	**without lab(s): 0.5 day**

OBJECTIVES

1. **Describe** two lines of evidence for Earth's rotation.
2. **Explain** how the change in apparent positions of constellations provides evidence of Earth's rotation and revolution around the sun.
3. **Summarize** how Earth's rotation and revolution provide a basis for measuring time.
4. **Explain** how the tilt of Earth's axis and Earth's movement cause seasons.

NATIONAL SCIENCE EDUCATION STANDARDS

SAI 2c: Scientists rely on technology to enhance the gathering and manipulation of data. New techniques and tools provide new evidence to guide inquiry and new methods to gather data, thereby contributing to the advance of science. The accuracy and precision of the data, and therefore the quality of the exploration, depends on the technology used.

SAI 2d: Mathematics is essential in scientific inquiry. Mathematical tools and models guide and improve the posing of questions, gathering data, constructing explanations and communicating results.

UCP: Unifying Concepts and Processes

FOCUS *(5 minutes)*

__ **Overview, TE** Review the objectives listed in the Student Edition. **(GENERAL)**

__ **PowerPoint® Resources** Use the customizable presentation to enhance your lecture.

__ **Bellringer, TE** Students answer questions about the rising and setting of the sun. **(GENERAL)**

__ **Bellringer Transparency** Use this transparency as students enter the classroom and find their seats.

MOTIVATE *(10 minutes)*

__ **Activity, Role-Playing, TE** Using a flashlight, students simulate Earth's movement around the sun. **(GENERAL)**

TEACH *(65 minutes)*

__ **Directed Reading, CRF** These worksheets reinforce basic concepts and vocabulary presented in the section. **(BASIC)**

KEY SE = Student Edition **TE** = Teacher Edition **CRF** = Chapter Resource File

Lesson Plan

__ **Quick Lab, Modeling a Pendulum, SE** Students use yo-yos to see how pendulums swing. **(GENERAL)**

__ **Datasheet for Quick Lab, Modeling a Pendulum, CRF** Students use the datasheet to complete the Quick Lab. **(GENERAL)**

__ **Demonstration, Coriolis Effect, TE** Students observe and describe a marble rolling across a spinning turntable. **(BASIC)**

__ **Skill Builder, Vocabulary, TE** Students review the origins of the words *aphelion* and *perihelion*. **(GENERAL)**

__ **Discussion, Comprehension Check, TE** Students discuss why ancient cultures wanted to remember the locations of stars. **(BASIC)**

__ **Activity, Star Viewers, TE** Students make star viewers using cardboard tubes and construction paper. **(GENERAL)**

__ **Graphic Organizer, Venn Diagram, SE** Use the graphic organizer to help students develop their analytical skills. **(GENERAL)**

__ **History Connection, It's About Time, TE** Students research different calendar systems. **(GENERAL)**

__ **Reading Skill Builder, Reading Hint, TE** Students practice "skimming and scanning" and brainstorm ways to get the most out of this technique. **(BASIC)**

__ **Internet Activity, International Zones, TE** Students research the evolution of time zones. **(GENERAL)**

__ **Internet Activity, International Zones, CRF** This worksheet directs student research for the topic presented in the Teacher Edition. **(GENERAL)**

__ **Using the Figure, Time Zone Zigzags, TE** Students discuss what happens as you move west from one time zone to the next. **(GENERAL)**

__ **Math Connection, What Time Is It? TE** Students use the map of time zones to determine what time it would be in different parts of the world. **(GENERAL)**

__ **Debate, Daylight Savings Time, TE** Students debate the merits of daylight savings time as a public policy. **(GENERAL)**

__ **Quick Lab, The Angle of the Sun's Rays, SE** Students use flashlights to illustrate the angle that the sun's rays hit Earth. **(GENERAL)**

__ **Datasheet for Quick Lab, The Angle of the Sun's Rays, CRF** Students use the datasheet to complete the Quick Lab. **(GENERAL)**

__ **Inquiry Lab, Earth-Sun Motion, SE** Students examine how shadow sticks can measure time. **(GENERAL)**

__ **Datasheet for Chapter Lab, Inquiry Lab, Earth-Sun Motion, CRF** Students use the datasheet to complete the Chapter Lab. **(GENERAL)**

__ **Science and Technology, Landsat Maps of Earth, SE** This article explains Landsat technology and discusses its numerous uses for Earth scientists.

__ **Internet Activity, Landsat Images, TE** In groups, students research the various applications of Landsat technologies. **(GENERAL)**

KEY **SE** = Student Edition **TE** = Teacher Edition **CRF** = Chapter Resource File

__ **Internet Activity, Landsat Images, CRF** This worksheet directs student research for the topic presented in the Teacher Edition. **(GENERAL)**

__ **Interactive Tutor CD-ROM, Gravity, Orbits, Motion, and Forces** Students practice what they have learned using this interactive activity. **(GENERAL)**

CLOSE *(10 minutes)*

__ **Section Review, SE** Students answer end-of-section vocabulary, key ideas, critical thinking, and concept mapping questions. **(GENERAL)**

__ **Reteaching, Evidence, TE** Students create a cause and effect organizer to understand the evidence for Earth's motions. **(BASIC)**

__ **Quiz, TE** Students answer 3 questions about this section. **(GENERAL)**

__ **Section Quiz, CRF** Use this quiz to assess students' understanding of the section. **(GENERAL) Also in Spanish**

__ **Spanish Section Quiz, Spanish Assessments** Use this quiz to assess students' understanding of the section. **(GENERAL)**

__ **Alternative Assessment, Science Fiction Stories, TE** Students write stories using the evidence for Earth's motions to solve a mystery. **(ADVANCED)**

KEY **SE** = Student Edition **TE** = Teacher Edition **CRF** = Chapter Resource File

End of Chapter Review and Assessment

PACING

Regular Schedule:	**with lab(s): N/A**	**without lab(s): 2 days**
Block Schedule:	**with lab(s): N/A**	**without lab(s): 1 day**

__ **Chapter Review, SE** Students answer end-of-chapter vocabulary, key ideas, critical thinking, math, writing, and interpreting graphics questions. **(GENERAL)**

__ **Alternative Assessment, Astronomy Picture Books, TE** Students make books showing how the universe is studied and how Earth's motion relates to time measurement. **(GENERAL)**

__ **Standardized Test Prep, SE** Have students complete the standardized test prep to help them prepare for standardized tests. **(GENERAL)**

__ **Holt PuzzlePro®** Use this software to create crossword puzzles and word searches that make learning vocabulary fun. **(GENERAL)**

__ **Concept Review, CRF** Use this worksheet to assess student's understanding of the main concepts presented in the chapter. **(GENERAL)**

__ **Critical Thinking, CRF** Students use this worksheet to develop and practice critical thinking skills. **(ADVANCED)**

__ **Math Skills, CRF** This worksheet helps reinforce students' understanding of order of operations. **(GENERAL)**

__ **Graphing Skills, CRF** This worksheet helps reinforce students' understanding of line graphs. **(GENERAL)**

__ **Concept Review, Study Guide, Studying Space** Use this worksheet to assess student's understanding of the main concepts presented in the chapter. **(GENERAL)**

__ **Concept Review, Spanish Study Guide, Studying Space** This worksheet helps reinforce English-language learners' understanding of the main concepts presented in the chapter. **(GENERAL)**

__ **Chapter Test A, CRF** Assign questions from the appropriate test for chapter assessment. **(GENERAL)**

__ **Chapter Test B, CRF** Assign questions from the appropriate test for chapter assessment. **(ADVANCED)**

__ **Spanish Chapter Test A, Spanish Assessments** Assign questions from the appropriate test for chapter assessment. **(GENERAL)**

__ **Test Generator, One-Stop Planner** Create a customized homework assignment, quiz, or test using the HRW Test Generator program. **(GENERAL)**

__ **Test Item Listing, CRF** Use the Test Item Listing to identify questions for a customized homework, quiz, or test.

__ **PowerPoint® Resources** Use the customizable presentation to help students practice standardized test-taking skills.

KEY SE = Student Edition TE = Teacher Edition CRF = Chapter Resource File

OTHER RESOURCE OPTIONS

__ **NOVA Video, "Runaway Universe"** Join two competing teams in their quest to discover the secrets of the stars and the ultimate fate of the universe.

__ **Brain Food Video Quiz** Use this video resource to assess student's progress. **(GENERAL)**

KEY **SE** = Student Edition **TE** = Teacher Edition **CRF** = Chapter Resource File

Chapter 27 Planets of the Solar System

Chapter Opener

__ **Pre-Reading Activity, Three-Panel Flip Chart, SE**

__ **Using the Figure, Martian Canyons, TE (GENERAL)**

__ **Long-Term Project, Planetary Motions, SE** Students study planetary motion by observing Mars twice a month for eight months. **(ADVANCED)**

__ **Long-Term Project, Planetary Motions, Long-term Projects** Students use this datasheet to complete the Long-term Project. **(ADVANCED)**

__ **Long-Term Project, Positions of Sunrise and Sunset, SE** Students measure changes in the direction of sunrise and sunset for 8 or 9 months. **(ADVANCED)**

__ **Long-Term Project, Positions of Sunrise and Sunset, Long-term Projects** Students use this datasheet to complete the Long-Term Project. **(ADVANCED)**

Section 1 Formation of the Solar System

PACING

Regular Schedule:	**with lab(s): N/A**	**without lab(s): 2 days**
Block Schedule:	**with lab(s): N/A**	**without lab(s): 1 day**

OBJECTIVES

1. **Explain** the nebular hypothesis of the origin of the solar system.
2. **Describe** how the planets formed.
3. **Describe** the formation of the land, the atmosphere, and the oceans of Earth.

NATIONAL SCIENCE EDUCATION STANDARDS

ES 3a: The sun, the earth, and the rest of the solar system formed from a nebular cloud of dust and gas 4.6 billion years ago. The early earth was very different from the planet we live on today.

UCP 1: The natural and designed world is complex; it is too large and complicated to investigate and comprehend all at once. Scientists and students learn to define small portions for the convenience of investigation. The units of investigation can be referred to as "systems." A system is an organized group of related objects or components that form a whole. Systems can consist, for example, of organisms, machines, fundamental particles, galaxies, ideas, numbers, transportation, and education. Systems have boundaries, components, resources flow (input and output), and feedback.

KEY **SE** = Student Edition **TE** = Teacher Edition **CRF** = Chapter Resource File

Lesson Plan

FOCUS *(5 minutes)*

__ **Overview, TE** Review the objectives listed in the Student Edition. **(GENERAL)**

__ **PowerPoint® Resources** Use the customizable presentation to enhance your lecture.

__ **Bellringer, TE** Have students write a list of all of the objects they think make up Earth's solar system. **(GENERAL)**

__ **Bellringer Transparency** Use this transparency as students enter the classroom and find their seats.

MOTIVATE *(10 minutes)*

__ **Group Activity, Spinning Nebula, TE** In groups, students create solar nebular models. **(GENERAL)**

TEACH *(20 minutes)*

__ **Directed Reading, CRF** These worksheets reinforce basic concepts and vocabulary presented in the section. **(BASIC)**

__ **Using the Figure, Discussion, TE** Students use the figures to discuss solar nebula and the formation of planets. **(GENERAL)**

__ **Reading Skill Builder, Reading Organizer, TE** Students are encouraged to take notes while reading this section. **(BASIC)**

__ **Teaching Transparency 136, The Nebular Model of the Formation of the Solar System with Worksheet** Use this Teaching Transparency and the accompanying worksheet to support your classroom discussion of the nebular model of the formation of the solar system.

__ **Group Activity, Skit, TE** Students perform skits about the formation of the solar system. **(GENERAL)**

__ **Quick Lab, Water Planetesimals, SE** Students use a medicine dropper, water, and wax paper to construct a water model of the formation of protoplanets. **(GENERAL)**

__ **Datasheet for Quick Lab, Planetesimals, CRF** Students use the datasheet to complete the Quick Lab. **(GENERAL)**

__ **Graphic Organizer, Chain-of-Events Chart, SE** Use the graphic organizer to help students develop their analytical skills. **(GENERAL)**

__ **Teaching Transparency 137, Differentiation of Earth and Formation of Earth's Atmosphere with Worksheet** Use this Teaching Transparency and the accompanying worksheet to support your classroom discussion of the differentiation of Earth and formation of Earth's atmosphere.

__ **Interactive Tutor CD-ROM, Our Sun and Origin of the Solar System** Students practice what they have learned using this interactive activity. **(GENERAL)**

CLOSE *(10 minutes)*

__ **Section Review, SE** Students answer end-of-section vocabulary, key ideas, critical thinking, and concept mapping questions. **(GENERAL)**

KEY **SE** = Student Edition **TE** = Teacher Edition **CRF** = Chapter Resource File

__ **Homework, Earth Spheres Posters, TE** Students create posters about one of Earth's spheres. **(GENERAL)**

__ **Reteaching, Flashcards, TE** Students make flashcards of the stages of Earth's formation. **(BASIC)**

__ **Quiz, TE** Students answer 3 questions about this section. **(GENERAL)**

__ **Section Quiz, CRF** Use this quiz to assess students' understanding of the section. **(GENERAL) Also in Spanish**

__ **Spanish Section Quiz, Spanish Assessments** Use this quiz to assess students' understanding of the section. **(GENERAL)**

__ **Alternative Assessment, Cosmic Cartoons, TE** Students draw cartoons illustrating the formation of the sun and planets from the solar nebula. **(ADVANCED)**

KEY **SE** = Student Edition **TE** = Teacher Edition **CRF** = Chapter Resource File

Lesson Plan

Section 2 Models of the Solar System

PACING

Regular Schedule:	**with lab(s): N/A**	**without lab(s): 2 days**
Block Schedule:	**with lab(s): N/A**	**without lab(s): 1 day**

OBJECTIVES

1. **Compare** the models of the universe developed by Ptolemy and Copernicus.
2. **Summarize** Kepler's three laws of planetary motion.
3. **Describe** how Newton explained Kepler's laws of motion.

NATIONAL SCIENCE EDUCATION STANDARDS

ST 1e: Communicate the problem, process, and solution. Students should present their results to students, teachers, and others in a variety of ways, such as orally, in writing, and in other forms--including models, diagrams, and demonstrations.

UCP 2: Evidence consists of observations and data on which to base scientific explanations. Using evidence to understand interactions allows individuals to predict changes in natural and designed systems. Models are tentative schemes or structures that correspond to real objects, events, or classes of events, and that have explanatory power. Models help scientists and engineers understand how things work. Models take many forms, including physical objects, plans, mental constructs, mathematical equations, and computer simulations.

FOCUS *(5 minutes)*

__ **Overview, TE** Review the objectives listed in the Student Edition. **(GENERAL)**

__ **PowerPoint® Resources** Use the customizable presentation to enhance your lecture.

__ **Bellringer, TE** Ask students to draw a diagram of the solar system and to label all of the parts of the diagram. **(GENERAL)**

__ **Bellringer Transparency** Use this transparency as students enter the classroom and find their seats.

MOTIVATE *(10 minutes)*

__ **Activity, Role Play, TE** Students perform an activity to help them understand how the planets appear to move backward. **(GENERAL)**

KEY **SE** = Student Edition **TE** = Teacher Edition **CRF** = Chapter Resource File

TEACH *(20 minutes)*

__ **Directed Reading, CRF** These worksheets reinforce basic concepts and vocabulary presented in the section. **(BASIC)**

__ **Quick Lab, Ellipses, SE** Students draw ellipses to review how planets orbit the sun. **(GENERAL)**

__ **Datasheet for Quick Lab, Ellipses, CRF** Students use the datasheet to complete the Quick Lab. **(GENERAL)**

__ **Skill Builder, Vocabulary, TE** Students review the origin of the words *ellipse* and *eccentricity*. **(GENERAL)**

__ **Teaching Transparency 138, Kepler's Law of Equal Area with Worksheet** Use this Teaching Transparency and the accompanying worksheet to support your classroom discussion of Kepler's law of equal area.

__ **Career Focus, Astronomer, SE** An astronomer describes how she uses different telescopes to study other galaxies.

__ **Group Activity, Interview an Astronomer, TE** Students interview an astronomer from a planetarium, space museum, or university and present a report. **(GENERAL)**

__ **Interactive Tutor CD-ROM, Gravity, Orbits, Motion, and Forces** Students practice what they have learned using this interactive activity. **(GENERAL)**

CLOSE *(10 minutes)*

__ **Section Review, SE** Students answer end-of-section vocabulary, key ideas, critical thinking, and concept mapping questions. **(GENERAL)**

__ **Reteaching, TE** Students create a Venn diagram to compare the geocentric and heliocentric models of the solar system. **(BASIC)**

__ **Quiz, TE** Students answer 3 questions about this section. **(GENERAL)**

__ **Section Quiz, CRF** Use this quiz to assess students' understanding of the section. **(GENERAL) Also in Spanish**

__ **Spanish Section Quiz, Spanish Assessments** Use this quiz to assess students' understanding of the section. **(GENERAL)**

__ **Alternative Assessment, TE** Students create an advertising "jingle" for one of Kepler's laws of motion. **(ADVANCED)**

KEY **SE** = Student Edition **TE** = Teacher Edition **CRF** = Chapter Resource File

Lesson Plan

Section 3 The Inner Planets

PACING

Regular Schedule:	**with lab(s): 2 days**	**without lab(s): 1 day**
Block Schedule:	**with lab(s): 1 day**	**without lab(s): 0.5 day**

OBJECTIVES

1. **Identify** the basic characteristics of the inner planets.
2. **Compare** the characteristics of the inner planets.
3. **Summarize** the features that allow Earth to sustain life.

NATIONAL SCIENCE EDUCATION STANDARDS

SAI 1a: Identify questions and concepts that guide scientific investigations. Students should formulate a testable hypothesis and demonstrate the logical connections between the scientific concepts guiding a hypothesis and the design of an experiment. They should demonstrate appropriate procedures, a knowledge base, and conceptual understanding of scientific investigations.

ST 1d: Evaluate the solution and its consequences. Students should test any solution against the needs and criteria it was designed to meet. At this stage, new criteria not originally considered may be reviewed.

ST 1e: Communicate the problem, process, and solution. Students should present their results to students, teachers, and others in a variety of ways, such as orally, in writing, and in other forms--including models, diagrams, and demonstrations.

FOCUS *(5 minutes)*

__ **Overview, TE** Review the objectives listed in the Student Edition. **(GENERAL)**

__ **PowerPoint® Resources** Use the customizable presentation to enhance your lecture.

__ **Bellringer, TE** Have students write a brief description of the characteristics of Earth. **(GENERAL)**

__ **Bellringer Transparency** Use this transparency as students enter the classroom and find their seats.

MOTIVATE *(10 minutes)*

__ **Discussion, Terrestrial Planets, TE** Discuss with students why the terrestrial planets have unique characteristics. **(GENERAL)**

KEY **SE** = Student Edition **TE** = Teacher Edition **CRF** = Chapter Resource File

Lesson Plan

TEACH *(65 minutes)*

__ **Directed Reading, CRF** These worksheets reinforce basic concepts and vocabulary presented in the section. **(BASIC)**

__ **Demonstration, Global Greenhouse, TE** Students create models of the atmospheres of Earth and Venus. **(GENERAL)**

__ **Using the Figure, Computer Generated Image, TE** Students use the image of the Maat Mons volcano to review the topography of Venus. **(BASIC)**

__ **Discussion, Radar Revelations, TE** Students discuss the surface features of Venus that have been revealed by radar. **(GENERAL)**

__ **Group Activity, Remote Sensing, TE** In groups, students model how radar-imaging probes map a hidden surface. **(GENERAL)**

__ **Reading Skill Builder, Paired Summarizing, TE** Students read the text in pairs and take turns summarizing what they have read. **(BASIC)**

__ **Skill Builder, Writing, TE** Students write a short story or poem set on an inner planet other than Earth. **(GENERAL)**

__ **Internet Activity, Life on Mars? TE** Students research the evidence for life on Mars. **(ADVANCED)**

__ **Internet Activity, Life on Mars? CRF** This worksheet directs student research for the topic presented in the Teacher Edition. **(ADVANCED)**

__ **Making Models Lab, Crater Analysis, SE** Students make model craters see how speed and projectile angle affect crater formation. **(GENERAL)**

__ **Datasheet for Chapter Lab, Making Models Lab, Crater Analysis, CRF** Students use the datasheet to complete the Chapter Lab. **(GENERAL)**

__ **Maps in Action, MOLA Map of Mars, SE** Students enhance their map-reading skills by answering questions about a map showing the relative elevation of surface features on Mar's surface. **(GENERAL)**

__ **Teaching Transparency 139, MOLA Map of Mars with Worksheet** Use this Teaching Transparency and the accompanying worksheet to help students perform the Maps in Action activity.

__ **Group Activity, Relief Models** Students create 3-D contour maps based on the information in the Martian relief map. **(GENERAL)**

__ **Inquiry Lab, Probing for Information, CRF** Students pretend to be scientists from another galaxy designing a mission and probe to study the Milky Way. **(GENERAL)**

__ **Interactive Tutor CD-ROM, The Inner Planets** Students practice what they have learned using this interactive activity. **(GENERAL)**

CLOSE *(10 minutes)*

__ **Section Review, SE** Students answer end-of-section vocabulary, key ideas, critical thinking, and concept mapping questions. **(GENERAL)**

__ **Reteaching, Same and Different, TE** Students write questions about the inner planets and quiz each other. **(BASIC)**

KEY SE = Student Edition TE = Teacher Edition CRF = Chapter Resource File

Lesson Plan

__ **Quiz, TE** Students answer 3 questions about this section. **(GENERAL)**

__ **Section Quiz, CRF** Use this quiz to assess students' understanding of the section. **(GENERAL) Also in Spanish**

__ **Spanish Section Quiz, Spanish Assessments** Use this quiz to assess students' understanding of the section. **(GENERAL)**

__ **Alternative Assessment, Planetary Base, TE** In groups, students plan and design a permanent colony or base on an inner planet other than Earth. **(ADVANCED)**

KEY **SE** = Student Edition **TE** = Teacher Edition **CRF** = Chapter Resource File

Lesson Plan

Section 4 The Outer Planets

PACING

Regular Schedule: **with lab(s): N/A** **without lab(s): 2 days**
Block Schedule: **with lab(s): N/A** **without lab(s): 1 day**

OBJECTIVES

1. **Identify** the basic characteristics that make the outer planets different from terrestrial planets.
2. **Compare** the characteristics of the outer planets.
3. **Explain** why Pluto is different from the other eight planets.

NATIONAL SCIENCE EDUCATION STANDARDS

ST 1e: Communicate the problem, process, and solution. Students should present their results to students, teachers, and others in a variety of ways, such as orally, in writing, and in other forms--including models, diagrams, and demonstrations.

FOCUS *(5 minutes)*

__ **Overview, TE** Review the objectives listed in the Student Edition. **(GENERAL)**

__ **PowerPoint® Resources** Use the customizable presentation to enhance your lecture.

__ **Bellringer, TE** Ask students to list what they know about the properties of gases. **(GENERAL)**

__ **Bellringer Transparency** Use this transparency as students enter the classroom and find their seats.

MOTIVATE *(10 minutes)*

__ **Demonstration, Play Ball, TE** Students compare the relative sizes of the planets using marbles and balls. **(GENERAL)**

TEACH *(20 minutes)*

__ **Directed Reading, CRF** These worksheets reinforce basic concepts and vocabulary presented in the section. **(BASIC)**

__ **Meteorology Connection, Lightning Discharge Model, TE** Students build models to demonstrate how electric charges build up in the clouds. **(GENERAL)**

__ **Using the Figure, Discussion, TE** Students use the figure of Jupiter to describe Jupiter's internal structure. **(GENERAL)**

__ **Demonstration, Gas Storms, TE** Students use a glass jar, water, vinegar, dish soap, and glitter or food coloring to demonstrate storms on Jupiter. **(GENERAL)**

KEY **SE** = Student Edition **TE** = Teacher Edition **CRF** = Chapter Resource File

__ **Discussion, What If…? TE** Students discuss how the Solar System might be different if Jupiter were a star with a larger mass than it actually has. **(GENERAL)**

__ **Discussion, Saturn's Rings, TE** Students discuss Saturn's rings. **(GENERAL)**

__ **Technology Connection, Cassini Mission, TE** Students research how the *Huygens* scientific probe and other instruments perform their jobs. **(ADVANCED)**

__ **Demonstration, Radically Tilted Planet, TE** Students discuss possible causes for Uranus's tilt. **(GENERAL)**

__ **Discussion, Composition and Structure, TE** Students discuss the atmosphere of Uranus. **(BASIC)**

__ **History Connection, Caroline Herschel, TE** Students research the astronomer Caroline Herschel and present a report of their findings. **(ADVANCED)**

__ **Discussion, Neptune's Weather, TE** Students answer questions about the weather on Neptune. **(GENERAL)**

__ **Group Activity, To Be or Not to Be a Planet, TE** Students debate whether Pluto should be considered a planet or a Kuiper Belt object. **(ADVANCED)**

__ **Internet Activity, Gravitational Microlensing, TE** Students research gravitational microlensing. **(ADVANCED)**

__ **Internet Activity, Gravitational Microlensing, CRF** This worksheet directs student research for the topic presented in the Teacher Edition. **(ADVANCED)**

__ **Making Models Lab, It's a Long Way to Pluto, CRF** Students use astronomical units to create a scale model of the Solar System. **(GENERAL)**

__ **Interactive Tutor CD-ROM, The Outer Planets** Students practice what they have learned using this interactive activity. **(GENERAL)**

CLOSE *(10 minutes)*

__ **Section Review, SE** Students answer end-of-section vocabulary, key ideas, critical thinking, and **interpreting graphics** questions. **(GENERAL)**

__ **Reteaching, Concept Webs, TE** Students work in pairs to create concept webs of the gas giants. **(BASIC)**

__ **Quiz, TE** Students answer 2 questions about this section. **(GENERAL)**

__ **Section Quiz, CRF** Use this quiz to assess students' understanding of the section. **(GENERAL) Also in Spanish**

__ **Spanish Section Quiz, Spanish Assessments** Use this quiz to assess students' understanding of the section. **(GENERAL)**

__ **Alternative Assessment, Outer Planets Museum, TE** Students create a mini space museum about the outer planets. **(GENERAL)**

KEY **SE** = Student Edition **TE** = Teacher Edition **CRF** = Chapter Resource File

End of Chapter Review and Assessment

PACING

Regular Schedule: **with lab(s): N/A** **without lab(s): 2 days**
Block Schedule: **with lab(s): N/A** **without lab(s): 1 day**

__ **Chapter Review, SE** Students answer end-of-chapter vocabulary, key ideas, critical thinking, math, writing, and interpreting graphics questions. **(GENERAL)**

__ **Alternative Assessment, Solar System Highlights, TE** Students work in groups to design fantasy Space Explorer Guides for the solar system. **(GENERAL)**

__ **Standardized Test Prep, SE** Have students complete the standardized test prep to help them prepare for standardized tests. **(GENERAL)**

__ **Holt PuzzlePro®** Use this software to create crossword puzzles and word searches that make learning vocabulary fun. **(GENERAL)**

__ **Concept Review, CRF** Use this worksheet to assess student's understanding of the main concepts presented in the chapter. **(GENERAL)**

__ **Critical Thinking, CRF** Students use this worksheet to develop and practice critical thinking skills. **(ADVANCED)**

__ **Math Skills, CRF** This worksheet helps reinforce students' understanding of exponents. **(GENERAL)**

__ **Graphing Skills, CRF** This worksheet helps reinforce students' understanding of bar graphs. **(GENERAL)**

__ **Concept Review, Study Guide, Planets of the Solar System** Use this worksheet to assess student's understanding of the main concepts presented in the chapter. **(GENERAL)**

__ **Concept Review, Spanish Study Guide, Planets of the Solar System** This worksheet helps reinforce English-language learners' understanding of the main concepts presented in the chapter. **(GENERAL)**

__ **Chapter Test A, CRF** Assign questions from the appropriate test for chapter assessment. **(GENERAL)**

__ **Chapter Test B, CRF** Assign questions from the appropriate test for chapter assessment. **(ADVANCED)**

__ **Spanish Chapter Test A, Spanish Assessments** Assign questions from the appropriate test for chapter assessment. **(GENERAL)**

__ **Test Generator, One-Stop Planner** Create a customized homework assignment, quiz, or test using the HRW Test Generator program. **(GENERAL)**

__ **Test Item Listing, CRF** Use the Test Item Listing to identify questions for a customized homework, quiz, or test.

__ **PowerPoint® Resources** Use the customizable presentation to help students practice standardized test-taking skills.

KEY **SE** = Student Edition **TE** = Teacher Edition **CRF** = Chapter Resource File

OTHER RESOURCE OPTIONS

__ **HRW Earth Science Video, Segment 18, "The Solar System"** In this segment the birth of the Solar System is examined, including everything from the formation of planets to the composition of comets and asteroids. **(GENERAL)**

__ **Brain Food Video Quiz** Use this video resource to assess student's progress. **(GENERAL)**

KEY **SE** = Student Edition **TE** = Teacher Edition **CRF** = Chapter Resource File

Lesson Plan

Chapter 28 Minor Bodies of the Solar System

Chapter Opener

__ **Pre-Reading Activity, Booklet, SE**

__ **Using the Figure, Asteroids, TE (GENERAL)**

__ **Long-Term Project, Apparent Motions of the Moon, Long-Term Projects** Students make daily observations of the moon to determine the path and phases of the moon. **(ADVANCED)**

__ **Long-Term Project, Tides at the Shoreline, Long-Term Projects** Students observe the changes and patterns of the tidal range of their area over a period of two months. **(ADVANCED)**

Section 1 Earth's Moon

PACING

Regular Schedule:	**with lab(s): N/A**	**without lab(s): 2 days**
Block Schedule:	**with lab(s): N/A**	**without lab(s): 1 day**

OBJECTIVES

1. **List** four kinds of lunar surface features.
2. **Describe** the three layers of the moon.
3. **Summarize** the three stages by which the moon formed.

NATIONAL SCIENCE EDUCATION STANDARDS

ES 3a: The sun, the earth, and the rest of the solar system formed from a nebular cloud of dust and gas 4.6 billion years ago. The early earth was very different from the planet we live on today.

HNS 2c: Because all scientific ideas depend on experimental and observational confirmation, all scientific knowledge is, in principle, subject to change as new evidence becomes available. The core ideas of science such as the conservation of energy or the laws of motion have been subjected to a wide variety of confirmations and are therefore unlikely to change in the areas in which they have been tested. In areas where data or understanding are incomplete, such as the details of human evolution or questions surrounding global warming, new data may well lead to changes in current ideas or resolve current conflicts. In situations where information is still fragmentary, it is normal for scientific ideas to be incomplete, but this is also where the opportunity for making advances may be greatest.

HNS 3c: Occasionally, there are advances in science and technology that have important and long-lasting effects on science and society. Examples of such advances include the following: Copernican revolution; Newtonian mechanics; Relativity; Geologic time scale;

KEY **SE** = Student Edition **TE** = Teacher Edition **CRF** = Chapter Resource File

Plate tectonics; Atomic theory; Nuclear physics; Biological evolution; Germ theory; Industrial revolution; Molecular biology; Information and communication; Quantum theory; Galactic universe; Medical and health technology.

PS 4b: Gravitation is a universal force that each mass exerts on any other mass. The strength of the gravitational attractive force between two masses is proportional to the masses and inversely proportional to the square of the distance between them.

SAI 2c: Scientists rely on technology to enhance the gathering and manipulation of data. New techniques and tools provide new evidence to guide inquiry and new methods to gather data, thereby contributing to the advance of science. The accuracy and precision of the data, and therefore the quality of the exploration, depends on the technology used.

SAI 2d: Mathematics is essential in scientific inquiry. Mathematical tools and models guide and improve the posing of questions, gathering data, constructing explanations and communicating results.

SAI 2f: Results of scientific inquiry—new knowledge and methods—emerge from different types of investigations and public communication among scientists. In communicating and defending the results of scientific inquiry, arguments must be logical and demonstrate connections between natural phenomena, investigations, and the historical body of scientific knowledge. In addition, the methods and procedures that scientists used to obtain evidence must be clearly reported to enhance opportunities for further investigation.

ST 2c: Creativity, imagination, and a good knowledge base are all required in the work of science and engineering.

UCP 3: Although most things are in the process of becoming different—changing—some properties of objects and processes are characterized by constancy, including the speed of light, the charge of an electron, and the total mass plus energy in the universe. Changes might occur, for example, in properties of materials, position of objects, motion, and form and function of systems. Interactions within and among systems result in change. Changes vary in rate, scale, and pattern, including trends and cycles.

FOCUS *(5 minutes)*

__ **Overview, TE** Review the objectives listed in the Student Edition. **(GENERAL)**

__ **PowerPoint® Resources** Use the customizable presentation to enhance your lecture.

__ **Bellringer, TE** Students write paragraphs that describe what they know about the moon. **(GENERAL)**

__ **Bellringer Transparency** Use this transparency as students enter the classroom and find their seats.

MOTIVATE *(10 minutes)*

__ **Group Activity, Moon Rescue, TE** In groups, students list supplies and materials that they would need to build a moon base for 100 people. **(GENERAL)**

KEY **SE** = Student Edition **TE** = Teacher Edition **CRF** = Chapter Resource File

Lesson Plan

TEACH *(20 minutes)*

__ **Directed Reading, CRF** These worksheets reinforce basic concepts and vocabulary presented in the section. **(BASIC)**

__ **Skill Builder, Math, TE** Students study a model of the Earth-Moon system to understand the enormous distances involved. **(GENERAL)**

__ **Using the Figure, Research, TE** Students research scientists who have had lunar features named after them. **(GENERAL)**

__ **Life Science Connection, Moon Trees, TE** Students research the locations of known moon trees in their state. **(GENERAL)**

__ **Discussion, Moon Rocks, TE** Students compare lunar rocks with rocks found on Earth. **(GENERAL)**

__ **Using the Figure, Footprints on the Moon TE** Students identify the regolith in the figure and explain why Neil Armstrong's footprints remain visible for a long time. **(GENERAL)**

__ **Quick Lab, Liquid and Solid Cores, SE** Students observe how objects with liquid and solid cores rotate using eggs, cans of soup, and solid food. **(GENERAL)**

__ **Datasheet for Quick Lab, Liquid and Solid Cores, CRF** Students use the datasheet to complete the Quick Lab. **(GENERAL)**

__ **Discussion, Comprehension Check, TE** Students discuss whether the interior of the moon is similar to Earth's interior. **(GENERAL)**

__ **Reading Skill Builder, Paired Summarizing, TE** Students read the text in pairs and take turns summarizing what they have read. **(BASIC)**

__ **Debate, Lunar Origins, TE** Students research and debate the moon's giant impact hypothesis and other competing hypotheses. **(ADVANCED)**

__ **Teaching Transparency 140, Formation of the Moon with Worksheet** Use this Teaching Transparency and the accompanying worksheet to support your classroom discussion of the formation of the moon.

__ **Maps in Action, Lunar Landing Sites, SE** Students enhance their map-reading skills by answering questions about a map showing lunar landing sites. **(GENERAL)**

__ **Teaching Transparency 145, Lunar Landing Sites with Worksheet** Use this Teaching Transparency and the accompanying worksheet to help students perform the Maps in Action activity.

__ **Internet Activity, Lunar Timeline, TE** Students research lunar missions. **(BASIC)**

__ **Internet Activity, Lunar Timeline, CRF** This worksheet directs student research for the topic presented in the Teacher Edition. **(BASIC)**

__ **Demonstration, Electrolysis of Water, TE** Students observe a demonstration of electrolysis to understand moon mining. **(GENERAL)**

__ **Science and Technology, Mining on the Moon, SE** This article discusses developing technologies to make mining on the moon possible.

__ **Making Models Lab, Crater Eraser, CRF** Students investigate a possible cause of hidden craters on active planets. **(GENERAL)**

KEY **SE** = Student Edition **TE** = Teacher Edition **CRF** = Chapter Resource File

CLOSE *(10 minutes)*

__ **Section Review, SE** Students answer end-of-section vocabulary, key ideas, critical thinking, and concept mapping questions. **(GENERAL)**

__ **Reteaching, Storyboards, TE** Students make storyboards illustrating the formation of the moon. **(BASIC)**

__ **Quiz, TE** Students answer 3 questions about this section. **(GENERAL)**

__ **Section Quiz, CRF** Use this quiz to assess students' understanding of the section. **(GENERAL) Also in Spanish**

__ **Spanish Section Quiz, Spanish Assessments** Use this quiz to assess students' understanding of the section. **(GENERAL)**

__ **Alternative Assessment, Lunar Picture Book, TE** Students create picture books explaining the topics in this section. **(GENERAL)**

KEY **SE** = Student Edition **TE** = Teacher Edition **CRF** = Chapter Resource File

Section 2 Movements of the Moon

PACING

Regular Schedule:	**with lab(s): N/A**	**without lab(s): 2 days**
Block Schedule:	**with lab(s): N/A**	**without lab(s): 1 day**

OBJECTIVES

1. **Describe** the shape of the moon's orbit around Earth.
2. **Explain** why eclipses occur.
3. **Describe** the appearance of four phases of the moon.
4. **Explain** how the movements of the moon affect tides on Earth.

NATIONAL SCIENCE EDUCATION STANDARDS

HNS 2c: Because all scientific ideas depend on experimental and observational confirmation, all scientific knowledge is, in principle, subject to change as new evidence becomes available. The core ideas of science such as the conservation of energy or the laws of motion have been subjected to a wide variety of confirmations and are therefore unlikely to change in the areas in which they have been tested. In areas where data or understanding are incomplete, such as the details of human evolution or questions surrounding global warming, new data may well lead to changes in current ideas or resolve current conflicts. In situations where information is still fragmentary, it is normal for scientific ideas to be incomplete, but this is also where the opportunity for making advances may be greatest.

HNS 3c: Occasionally, there are advances in science and technology that have important and long-lasting effects on science and society. Examples of such advances include the following: Copernican revolution; Newtonian mechanics; Relativity; Geologic time scale; Plate tectonics; Atomic theory; Nuclear physics; Biological evolution; Germ theory; Industrial revolution; Molecular biology; Information and communication; Quantum theory; Galactic universe; Medical and health technology.

PS 4b: Gravitation is a universal force that each mass exerts on any other mass. The strength of the gravitational attractive force between two masses is proportional to the masses and inversely proportional to the square of the distance between them.

SAI 2c: Scientists rely on technology to enhance the gathering and manipulation of data. New techniques and tools provide new evidence to guide inquiry and new methods to gather data, thereby contributing to the advance of science. The accuracy and precision of the data, and therefore the quality of the exploration, depends on the technology used.

SAI 2d: Mathematics is essential in scientific inquiry. Mathematical tools and models guide and improve the posing of questions, gathering data, constructing explanations and communicating results.

KEY **SE** = Student Edition **TE** = Teacher Edition **CRF** = Chapter Resource File

SAI 2f: Results of scientific inquiry—new knowledge and methods—emerge from different types of investigations and public communication among scientists. In communicating and defending the results of scientific inquiry, arguments must be logical and demonstrate connections between natural phenomena, investigations, and the historical body of scientific knowledge. In addition, the methods and procedures that scientists used to obtain evidence must be clearly reported to enhance opportunities for further investigation.

ST 2c: Creativity, imagination, and a good knowledge base are all required in the work of science and engineering.

UCP 1: The natural and designed world is complex; it is too large and complicated to investigate and comprehend all at once. Scientists and students learn to define small portions for the convenience of investigation. The units of investigation can be referred to as "systems." A system is an organized group of related objects or components that form a whole. Systems can consist, for example, of organisms, machines, fundamental particles, galaxies, ideas, numbers, transportation, and education. Systems have boundaries, components, resources flow (input and output), and feedback.

UCP 3: Although most things are in the process of becoming different—changing—some properties of objects and processes are characterized by constancy, including the speed of light, the charge of an electron, and the total mass plus energy in the universe. Changes might occur, for example, in properties of materials, position of objects, motion, and form and function of systems. Interactions within and among systems result in change. Changes vary in rate, scale, and pattern, including trends and cycles.

FOCUS *(5 minutes)*

__ **Overview, TE** Review the objectives listed in the Student Edition. **(GENERAL)**

__ **PowerPoint® Resources** Use the customizable presentation to enhance your lecture.

__ **Bellringer, TE** Students create a series of drawings that show how the moon's appearance changes over several weeks. **(GENERAL)**

__ **Bellringer Transparency** Use this transparency as students enter the classroom and find their seats.

MOTIVATE *(10 minutes)*

__ **Group Activity, Barycenter, TE** Students make a model showing how the Earth-moon system rotates around a center point. **(GENERAL)**

TEACH *(20 minutes)*

__ **Directed Reading, CRF** These worksheets reinforce basic concepts and vocabulary presented in the section. **(BASIC)**

__ **Teaching Transparency 141, The Earth-Moon System with Worksheet** Use this Teaching Transparency and the accompanying worksheet to support your classroom discussion of the Earth-moon system.

KEY **SE** = Student Edition **TE** = Teacher Edition **CRF** = Chapter Resource File

__ **Discussion, Lunar Orbit, TE** Students discuss how the moon's elliptical orbit causes the distance from Earth to the moon to vary. **(BASIC)**

__ **Demonstration, Synchronous Rotation, TE** Students use a swivel chair, a foam ball, and a dowel rod to demonstrate Earth's and the moon's orbits. **(GENERAL)**

__ **Teaching Transparency 142, Solar and Lunar Eclipses with Worksheet** Use this Teaching Transparency and the accompanying worksheet to support your classroom discussion of solar and lunar eclipses.

__ **Using the Figure, Solar Eclipse, TE** Students use the diagram of a solar eclipse to describe where people must be on Earth in order to see a solar eclipse. **(BASIC)**

__ **Activity, Eclipsed, TE** Students use coins or marbles to illustrate how the moon produces a total eclipse. **(GENERAL)**

__ **Quick Lab, Eclipses, SE** Students use modeling clay and a penlight to simulate an eclipse. **(GENERAL)**

__ **Datasheet for Quick Lab, Eclipses, CRF** Students use the datasheet to complete the Quick Lab. **(GENERAL)**

__ **Life Science Connection, Biological Rhythms, TE** Students research the effects of eclipses on the circadian rhythms of living things. **(GENERAL)**

__ **Group Activity, Many Moons, TE** Students use foam balls and a lamp to see how and why different parts of the moon light up. **(GENERAL)**

__ **Teaching Transparency 143, Phases of the Moon with Worksheet** Use this Teaching Transparency and the accompanying worksheet to support your classroom discussion of phases of the moon.

__ **Teaching Transparency 144, Causes of Tides with Worksheet** Use this Teaching Transparency and the accompanying worksheet to support your classroom discussion of the causes of tides.

__ **Skill Builder, Vocabulary, TE** Students examine roots of words that relate to the moon. **(GENERAL)**

__ **Inquiry Lab, Inconstant Moon, CRF** Students observe the moon for one month and record observations about how it changes. **(GENERAL)**

__ **Interactive Tutor CD-ROM, Waves and Tides** Students practice what they have learned using this interactive activity. **(GENERAL)**

CLOSE *(10 minutes)*

__ **Section Review, SE** Students answer end-of-section vocabulary, key ideas, critical thinking, and concept mapping questions. **(GENERAL)**

__ **Homework, Moon Watching, TE** Students keep a daily journal in which they record observations and patterns of the moon. **(GENERAL)**

__ **Reteaching, Lunar Flipbook, TE** Students create flipbooks showing the moon's phases. **(BASIC)**

__ **Quiz, TE** Students answer 2 questions about this section. **(GENERAL)**

KEY **SE** = Student Edition **TE** = Teacher Edition **CRF** = Chapter Resource File

__ **Section Quiz, CRF** Use this quiz to assess students' understanding of the section. **(GENERAL) Also in Spanish**

__ **Spanish Section Quiz, Spanish Assessments** Use this quiz to assess students' understanding of the section. **(GENERAL)**

__ **Alternative Assessment, System Models, TE** In groups, students make models of the moon and explain eclipses, tides, and the moon's phases. **(ADVANCED)**

KEY **SE** = Student Edition **TE** = Teacher Edition **CRF** = Chapter Resource File

Section 3 Satellites of Other Planets

PACING

Regular Schedule: **with lab(s): 2 days** **without lab(s): 1 day**
Block Schedule: **with lab(s): 1 day** **without lab(s): 0.5 day**

OBJECTIVES

1. **Compare** the characteristics of the two moons of Mars.
2. **Describe** how volcanoes were discovered on Io.
3. **Name** one distinguishing characteristic of each of the Galilean moons.
4. **Compare** the characteristics of the rings of Saturn with the rings of the other outer planets.

NATIONAL SCIENCE EDUCATION STANDARDS

HNS 2c: Because all scientific ideas depend on experimental and observational confirmation, all scientific knowledge is, in principle, subject to change as new evidence becomes available. The core ideas of science such as the conservation of energy or the laws of motion have been subjected to a wide variety of confirmations and are therefore unlikely to change in the areas in which they have been tested. In areas where data or understanding are incomplete, such as the details of human evolution or questions surrounding global warming, new data may well lead to changes in current ideas or resolve current conflicts. In situations where information is still fragmentary, it is normal for scientific ideas to be incomplete, but this is also where the opportunity for making advances may be greatest.

HNS 3c: Occasionally, there are advances in science and technology that have important and long-lasting effects on science and society. Examples of such advances include the following: Copernican revolution; Newtonian mechanics; Relativity; Geologic time scale; Plate tectonics; Atomic theory; Nuclear physics; Biological evolution; Germ theory; Industrial revolution; Molecular biology; Information and communication; Quantum theory; Galactic universe; Medical and health technology.

PS 4b: Gravitation is a universal force that each mass exerts on any other mass. The strength of the gravitational attractive force between two masses is proportional to the masses and inversely proportional to the square of the distance between them.

SAI 2c: Scientists rely on technology to enhance the gathering and manipulation of data. New techniques and tools provide new evidence to guide inquiry and new methods to gather data, thereby contributing to the advance of science. The accuracy and precision of the data, and therefore the quality of the exploration, depends on the technology used.

SAI 2d: Mathematics is essential in scientific inquiry. Mathematical tools and models guide and improve the posing of questions, gathering data, constructing explanations and communicating results.

KEY **SE** = Student Edition **TE** = Teacher Edition **CRF** = Chapter Resource File

Lesson Plan

SAI 2f: Results of scientific inquiry—new knowledge and methods—emerge from different types of investigations and public communication among scientists. In communicating and defending the results of scientific inquiry, arguments must be logical and demonstrate connections between natural phenomena, investigations, and the historical body of scientific knowledge. In addition, the methods and procedures that scientists used to obtain evidence must be clearly reported to enhance opportunities for further investigation.

ST 2c: Creativity, imagination, and a good knowledge base are all required in the work of science and engineering.

UCP 1: The natural and designed world is complex; it is too large and complicated to investigate and comprehend all at once. Scientists and students learn to define small portions for the convenience of investigation. The units of investigation can be referred to as "systems." A system is an organized group of related objects or components that form a whole. Systems can consist, for example, of organisms, machines, fundamental particles, galaxies, ideas, numbers, transportation, and education. Systems have boundaries, components, resources flow (input and output), and feedback.

UCP 3: Although most things are in the process of becoming different—changing—some properties of objects and processes are characterized by constancy, including the speed of light, the charge of an electron, and the total mass plus energy in the universe. Changes might occur, for example, in properties of materials, position of objects, motion, and form and function of systems. Interactions within and among systems result in change. Changes vary in rate, scale, and pattern, including trends and cycles.

FOCUS *(5 minutes)*

__ **Overview, TE** Review the objectives listed in the Student Edition. **(GENERAL)**

__ **PowerPoint® Resources** Use the customizable presentation to enhance your lecture.

__ **Bellringer, TE** Students write definitions of a moon. **(GENERAL)**

__ **Bellringer Transparency** Use this transparency as students enter the classroom and find their seats.

MOTIVATE *(10 minutes)*

__ **Discussion, Mythology Connection, TE** Students discuss the historical origins the planets' names. **(GENERAL)**

TEACH *(65 minutes)*

__ **Directed Reading, CRF** These worksheets reinforce basic concepts and vocabulary presented in the section. **(BASIC)**

__ **Internet Activity, Martian Moons, TE** Students research the moons of Mars. **(GENERAL)**

__ **Internet Activity, Martian Moons, CRF** This worksheet directs student research for the topic presented in the Teacher Edition. **(GENERAL)**

KEY **SE** = Student Edition **TE** = Teacher Edition **CRF** = Chapter Resource File

__ **Reading Skill Builder, Reading Organizer, TE** Students are encouraged to take notes while reading this section. **(BASIC)**

__ **Using the Figure, Geologic Pasts, TE** Students study the photos of Jupiter's moons and identify which moon is the oldest and the most geologically active. **(GENERAL)**

__ **Using the Figure, Ringmaster, TE** Students observe a model of the Saturn system and compare it to the inner moons of planetary rings. **(GENERAL)**

__ **Demonstration, Shepherd Moons, TE** Use the graphic organizer to help students develop their analytical skills. **(GENERAL)**

__ **Graphic Organizer, Comparison Table, SE** Use the graphic organizer to help students develop their analytical skills. **(GENERAL)**

__ **Meteorology Connection, Titan's Atmosphere, TE** Students research the Cassini mission to Saturn and Titan. **(GENERAL)**

__ **Activity, Planetary Rings, TE** Students use a Venn diagram to compare Saturn's rings with the rings of other planets. **(GENERAL)**

__ **Skills Practice Lab, Galilean Moons of Jupiter, SE** Students relate the orbits of Jupiter's moons to Kepler's third law of motion. **(GENERAL)**

__ **Datasheet for Chapter Lab, Skills Practice Lab, Galilean Moons of Jupiter, CRF** Students use the datasheet to complete the Chapter Lab. **(GENERAL)**

CLOSE *(10 minutes)*

__ **Section Review, SE** Students answer end-of-section vocabulary, key ideas, critical thinking, and concept mapping questions. **(GENERAL)**

__ **Reteaching, Peer Reviewing, TE** In small groups, students write questions about this section and quiz each other. **(BASIC)**

__ **Quiz, TE** Students answer 3 questions about this section. **(GENERAL)**

__ **Section Quiz, CRF** Use this quiz to assess students' understanding of the section. **(GENERAL) Also in Spanish**

__ **Spanish Section Quiz, Spanish Assessments** Use this quiz to assess students' understanding of the section. **(GENERAL)**

__ **Alternative Assessment, Satellite Bingo, TE** In groups, students create a bingo game using characteristics of the moons of the Solar System. **(GENERAL)**

KEY **SE** = Student Edition **TE** = Teacher Edition **CRF** = Chapter Resource File

Section 4 Asteroids, Comets, and Meteoroids

PACING

Regular Schedule: **with lab(s): N/A** **without lab(s): 2 days**
Block Schedule: **with lab(s): N/A** **without lab(s): 1 day**

OBJECTIVES

1. **Describe** the physical characteristics of asteroids and comets.
2. **Describe** where the Kuiper belt is located.
3. **Compare** meteoroids, meteorites, and meteors.
4. **Explain** the relationship between the Oort cloud and comets.

NATIONAL SCIENCE EDUCATION STANDARDS

HNS 2c: Because all scientific ideas depend on experimental and observational confirmation, all scientific knowledge is, in principle, subject to change as new evidence becomes available. The core ideas of science such as the conservation of energy or the laws of motion have been subjected to a wide variety of confirmations and are therefore unlikely to change in the areas in which they have been tested. In areas where data or understanding are incomplete, such as the details of human evolution or questions surrounding global warming, new data may well lead to changes in current ideas or resolve current conflicts. In situations where information is still fragmentary, it is normal for scientific ideas to be incomplete, but this is also where the opportunity for making advances may be greatest.

HNS 3c: Occasionally, there are advances in science and technology that have important and long-lasting effects on science and society. Examples of such advances include the following: Copernican revolution; Newtonian mechanics; Relativity; Geologic time scale; Plate tectonics; Atomic theory; Nuclear physics; Biological evolution; Germ theory; Industrial revolution; Molecular biology; Information and communication; Quantum theory; Galactic universe; Medical and health technology.

PS 4b: Gravitation is a universal force that each mass exerts on any other mass. The strength of the gravitational attractive force between two masses is proportional to the masses and inversely proportional to the square of the distance between them.

SAI 2c: Scientists rely on technology to enhance the gathering and manipulation of data. New techniques and tools provide new evidence to guide inquiry and new methods to gather data, thereby contributing to the advance of science. The accuracy and precision of the data, and therefore the quality of the exploration, depends on the technology used.

SAI 2d: Mathematics is essential in scientific inquiry. Mathematical tools and models guide and improve the posing of questions, gathering data, constructing explanations and communicating results.

KEY SE = Student Edition **TE** = Teacher Edition **CRF** = Chapter Resource File

SAI 2f: Results of scientific inquiry—new knowledge and methods—emerge from different types of investigations and public communication among scientists. In communicating and defending the results of scientific inquiry, arguments must be logical and demonstrate connections between natural phenomena, investigations, and the historical body of scientific knowledge. In addition, the methods and procedures that scientists used to obtain evidence must be clearly reported to enhance opportunities for further investigation.

UCP 1: The natural and designed world is complex; it is too large and complicated to investigate and comprehend all at once. Scientists and students learn to define small portions for the convenience of investigation. The units of investigation can be referred to as "systems." A system is an organized group of related objects or components that form a whole. Systems can consist, for example, of organisms, machines, fundamental particles, galaxies, ideas, numbers, transportation, and education. Systems have boundaries, components, resources flow (input and output), and feedback.

UCP 3: Although most things are in the process of becoming different—changing—some properties of objects and processes are characterized by constancy, including the speed of light, the charge of an electron, and the total mass plus energy in the universe. Changes might occur, for example, in properties of materials, position of objects, motion, and form and function of systems. Interactions within and among systems result in change. Changes vary in rate, scale, and pattern, including trends and cycles.

FOCUS *(5 minutes)*

__ **Overview, TE** Review the objectives listed in the Student Edition. **(GENERAL)**

__ **PowerPoint® Resources** Use the customizable presentation to enhance your lecture.

__ **Bellringer, TE** Students write descriptions of meteorites, comets, and asteroids. **(GENERAL)**

__ **Bellringer Transparency** Use this transparency as students enter the classroom and find their seats.

MOTIVATE *(10 minutes)*

__ **Discussion, Asteroid Motions, TE** Students describe where asteroids are located and how they move. **(GENERAL)**

TEACH *(20 minutes)*

__ **Directed Reading, CRF** These worksheets reinforce basic concepts and vocabulary presented in the section. **(BASIC)**

__ **Physics Connection, Unusual Orbits, TE** Students research asteroids with unusual orbits and create orbital diagrams for those asteroids. **(ADVANCED)**

__ **Using the Figure, Comparing Craters, TE** Students compare photos of craters with the Barringer Meteor Crater. **(GENERAL)**

__ **Using the Figure, Discussion, TE** Students discuss the Hale-Bopp comet and describe its basic features. **(GENERAL)**

KEY **SE** = Student Edition **TE** = Teacher Edition **CRF** = Chapter Resource File

__ **Using the Figure, Oort Cloud Distances, TE** Students visualize the vast distances of space by comparing the distance to a familiar landmark. **(GENERAL)**

__ **Internet Activity, Kuiper Belt Objects, TE** Students research objects in the Kuiper belt. **(GENERAL)**

__ **Internet Activity, Kuiper Belt Objects, CRF** This worksheet directs student research for the topic presented in the Teacher Edition. **(GENERAL)**

__ **Using the Figure, Leonid Meteor Shower, TE** Students research what constellation the Leonid shower was named for. **(GENERAL)**

__ **Interactive Tutor CD-ROM, Asteroids, Comets, and Meteors** Students practice what they have learned using this interactive activity. **(GENERAL)**

CLOSE *(10 minutes)*

__ **Section Review, SE** Students answer end-of-section vocabulary, key ideas, critical thinking, and concept mapping questions. **(GENERAL)**

__ **Reteaching, Crater Features, TE** Students examine the features of craters using flour, paint, and pebbles. **(BASIC)**

__ **Quiz, TE** Students answer 2 questions about this section. **(GENERAL)**

__ **Section Quiz, CRF** Use this quiz to assess students' understanding of the section. **(GENERAL) Also in Spanish**

__ **Spanish Section Quiz, Spanish Assessments** Use this quiz to assess students' understanding of the section. **(GENERAL)**

__ **Alternative Assessment, Science Nonficton, TE** Students research a minor body and write a description of a trip through the Solar System on that body. **(GENERAL)**

KEY **SE** = Student Edition **TE** = Teacher Edition **CRF** = Chapter Resource File

End of Chapter Review and Assessment

PACING

Regular Schedule: **with lab(s): N/A** **without lab(s): 2 days**
Block Schedule: **with lab(s): N/A** **without lab(s): 1 day**

__ **Chapter Review, SE** Students answer end-of-chapter vocabulary, key ideas, critical thinking, math, writing, and interpreting graphics questions. **(GENERAL)**

__ **Alternative Assessment, Poster Project, TE** Students create posters showing how studying objects like Earth's moon reveal the history of the Solar System. **(GENERAL)**

__ **Standardized Test Prep, SE** Have students complete the standardized test prep to help them prepare for standardized tests. **(GENERAL)**

__ **Holt PuzzlePro®** Use this software to create crossword puzzles and word searches that make learning vocabulary fun. **(GENERAL)**

__ **Concept Review, CRF** Use this worksheet to assess student's understanding of the main concepts presented in the chapter. **(GENERAL)**

__ **Critical Thinking, CRF** Students use this worksheet to develop and practice their critical thinking sills. **(ADVANCED)**

__ **Math Skills, CRF** This worksheet helps reinforce students' understanding of exponents. **(GENERAL)**

__ **Graphing Skills, CRF** This worksheet helps reinforce students' understanding of ternary diagrams. **(GENERAL)**

__ **Concept Review, Study Guide, Minor Bodies of the Solar System** Use this worksheet to assess student's understanding of the main concepts presented in the chapter. **(GENERAL)**

__ **Concept Review, Spanish Study Guide, Minor Bodies of the Solar System** This worksheet helps reinforce English-language learners' understanding of the main concepts presented in the chapter. **(GENERAL)**

__ **Chapter Test A, CRF** Assign questions from the appropriate test for chapter assessment. **(GENERAL)**

__ **Chapter Test B, CRF** Assign questions from the appropriate test for chapter assessment. **(ADVANCED)**

__ **Spanish Chapter Test A, Spanish Assessments** Assign questions from the appropriate test for chapter assessment. **(GENERAL)**

__ **Test Generator, One-Stop Planner** Create a customized homework assignment, quiz, or test using the HRW Test Generator program. **(GENERAL)**

__ **Test Item Listing, CRF** Use the Test Item Listing to identify questions for a customized homework, quiz, or test.

__ **PowerPoint® Resources** Use the customizable presentation to help students practice standardized test-taking skills.

KEY **SE** = Student Edition **TE** = Teacher Edition **CRF** = Chapter Resource File

OTHER RESOURCE OPTIONS

__ **HRW Earth Science Video, Segment 18, "The Solar System"** In this segment the birth of the Solar System is examined, including everything from the formation of planets to the composition of comets and asteroids. **(GENERAL)**

__ **HRW Earth Science Video, Segment 19, "The Earth's Moon"** This segment explores the various alignments of Earth, the moon, and the sun that give rise to the phases of the moon and that create lunar and solar eclipses. **(GENERAL)**

__ **Brain Food Video Quiz** Use this video resource to assess student's progress. **(GENERAL)**

KEY **SE** = Student Edition **TE** = Teacher Edition **CRF** = Chapter Resource File

Chapter 29 The Sun

Chapter Opener

__ **Pre-Reading Activity, Table Fold, SE**

__ **Using the Figure, The Explosive Sun, TE (GENERAL)**

__ **Long-Term Project, Positions of Sunrise and Sunset, SE** Students measure changes in the direction of sunrise and sunset for 8 or 9 months. **(ADVANCED)**

__ **Long-Term Project, Positions of Sunrise and Sunset, Long-term Projects** Students use this datasheet to complete the Long-Term Project. **(ADVANCED)**

Section 1 Structure of the Sun

PACING

Regular Schedule:	**with lab(s): 2 days**	**without lab(s): 1 day**
Block Schedule:	**with lab(s): 1 day**	**without lab(s): 0.5 day**

OBJECTIVES

1. **Explain** how the sun converts matter into energy in its core.
2. **Compare** the radiative and convection zones of the sun.
3. **Describe** the three layers of the sun's atmosphere.

NATIONAL SCIENCE EDUCATION STANDARDS

ES 4b: Early in the history of the universe, matter, primarily the light atoms hydrogen and helium, clumped together by gravitational attraction to form countless trillions of stars. Billions of galaxies, each of which is a gravitationally bound cluster of billions of stars, now form most of the visible mass in the universe.

ES 4c: Stars produce energy from nuclear reactions, primarily the fusion of hydrogen to form helium. These and other processes in stars have led to the formation of all the other elements.

PS 5c: Heat consists of random motion and the vibrations of atoms, molecules, and ions. The higher the temperature, the greater the atomic or molecular motion.

PS 5d: Everything tends to become less organized and less orderly over time. Thus, in all energy transfers, the overall effect is that the energy is spread out uniformly. Examples are the transfer of energy from hotter to cooler objects by conduction, radiation, or convection and the warming of our surroundings when we burn fuels.

KEY **SE** = Student Edition **TE** = Teacher Edition **CRF** = Chapter Resource File

Lesson Plan

FOCUS *(5 minutes)*

__ **Overview, TE** Review the objectives listed in the Student Edition. **(GENERAL)**

__ **PowerPoint® Resources** Use the customizable presentation to enhance your lecture.

__ **Bellringer, TE** Ask students "What is the sun made of?" and "What does the sun use as a fuel to produce energy?" **(GENERAL)**

__ **Bellringer Transparency** Use this transparency as students enter the classroom and find their seats.

MOTIVATE *(10 minutes)*

__ **Using the Figure, Discussion, TE** Students use the figure of dark lines that cross a spectrum to answer a question about hydrogen in the sun's atmosphere. **(GENERAL)**

TEACH *(65 minutes)*

__ **Directed Reading, CRF** These worksheets reinforce basic concepts and vocabulary presented in the section. **(BASIC)**

__ **Graphic Organizer, Chain-of-Events Chart, SE** Use the graphic organizer to help students develop their analytical skills. **(GENERAL)**

__ **Skill Builder, Math, TE** Students calculate a problem about fusion. **(ADVANCED)**

__ **Teaching Transparency 146, Nuclear Fusion with Worksheet** Use this Teaching Transparency and the accompanying worksheet to support your classroom discussion of nuclear fusion.

__ **Quick Lab, Modeling Fusion, SE** Students use coins to perform an experiment about fusion. **(GENERAL)**

__ **Datasheet for Quick Lab, Modeling Fusion, CRF** Students use the datasheet to complete the Quick Lab. **(GENERAL)**

__ **Using the Figure, Discussion, TE** Students use the figure showing a water tank to answer a question about subatomic particles. **(GENERAL)**

__ **History Connection, Reinventing Time and Space, TE** Students research the life and work of Albert Einstein. **(ADVANCED)**

__ **Quick Lab, The Size of Our Sun, SE** Students use paper and a compass to draw and measure Jupiter, Earth, and the Sun. **(GENERAL)**

__ **Datasheet for Quick Lab, The Size of Our sun, CRF** Students use the datasheet to complete the Quick Lab. **(GENERAL)**

__ **Reading Skill Builder, Reading Organizer, TE** Students are encouraged to take notes while reading this section. **(BASIC)**

__ **Teaching Transparency 147, The Sun's Interior with Worksheet** Use this Teaching Transparency and the accompanying worksheet to support your classroom discussion of the sun's interior.

__ **Skill Builder, Vocabulary, TE** Students review the roots of the names for the layers of the sun's atmosphere. **(GENERAL)**

KEY **SE** = Student Edition **TE** = Teacher Edition **CRF** = Chapter Resource File

__ **Skills Practice Lab, Energy of the Sun, SE** Students collect solar energy and estimate the total amount of energy produced by the sun. **(GENERAL)**

__ **Datasheet for Chapter Lab, Skills Practice Lab, Energy of the Sun, CRF** Students use the datasheet to complete the Chapter Lab. **(GENERAL)**

__ **Maps in Action, SXT Composite Image of the Sun, SE** Students answer questions about a map of a soft X-ray telescope image to enhance their map-reading skills. **(GENERAL)**

__ **Teaching Transparency 148, SXT Composite Image of the Sun with Worksheet** Use this Teaching Transparency and the accompanying worksheet to help students perform the Maps in Action activity.

__ **Internet Activity, SOHO Images, TE** Students research the SOHO Web site and construct a map of some aspect of solar activity. **(ADVANCED)**

__ **Internet Activity, SOHO Images, CRF** This worksheet directs student research for the topic presented in the Teacher Edition. **(ADVANCED)**

__ **Inquiry Lab, Solar Cooker, CRF** Students study methods of collecting solar energy in order to design and build a solar cooker. **(GENERAL)**

__ **Making Models Lab, Light Fingerprints, CRF** Students make model spectroscopes to study the spectra of different light sources. **(GENERAL)**

__ **Interactive Tutor CD-ROM, What's the Matter? (Atomic Theory and Nuclear Energy)** Students practice what they have learned using this interactive activity. **(GENERAL)**

__ **Interactive Tutor CD-ROM, Energy** Students practice what they have learned using this interactive activity. **(GENERAL)**

__ **Interactive Tutor CD-ROM, Light and Electromagnetic Energy** Students practice what they have learned using this interactive activity. **(GENERAL)**

CLOSE *(10 minutes)*

__ **Section Review, SE** Students answer end-of-section vocabulary, key ideas, critical thinking, and concept mapping questions. **(GENERAL)**

__ **Reteaching, Diagramming the Sun, TE** Students make a diagram of the sun's layers. **(BASIC)**

__ **Quiz, TE** Students answer 3 questions about this section. **(GENERAL)**

__ **Section Quiz, CRF** Use this quiz to assess students' understanding of the section. **(GENERAL) Also in Spanish**

__ **Spanish Section Quiz, Spanish Assessments** Use this quiz to assess students' understanding of the section. **(GENERAL)**

__ **Alternative Assessment, Solar Tales, TE** Students write stories about a hydrogen atom inside the sun. **(ADVANCED)**

KEY SE = Student Edition TE = Teacher Edition CRF = Chapter Resource File

Section 2 Solar Activity

PACING

Regular Schedule: **with lab(s): N/A** **without lab(s): 2 days**
Block Schedule: **with lab(s): N/A** **without lab(s): 1 day**

OBJECTIVES

1. **Explain** how sunspots are related to powerful magnetic fields on the sun.
2. **Compare** prominences, solar flares, and coronal mass ejections.
3. **Describe** how the solar wind can cause auroras on Earth.

NATIONAL SCIENCE EDUCATION STANDARDS

PS 4e: Electricity and magnetism are two aspects of a single electromagnetic force. Moving electric charges produce magnetic forces, and moving magnets produce electric forces. These effects help students to understand electric motors and generators.

FOCUS *(5 minutes)*

__ **Overview, TE** Review the objectives listed in the Student Edition. **(GENERAL)**

__ **PowerPoint® Resources** Use the customizable presentation to enhance your lecture.

__ **Bellringer, TE** Ask students, "What do you think the 'weather' in the sun's atmosphere is like? **(GENERAL)**

__ **Bellringer Transparency** Use this transparency as students enter the classroom and find their seats.

MOTIVATE *(10 minutes)*

__ **Discussion, Sunspots, TE** Students discuss why sunspots appear dark. **(GENERAL)**

TEACH *(20 minutes)*

__ **Directed Reading, CRF** These worksheets reinforce basic concepts and vocabulary presented in the section. **(BASIC)**

__ **Activity, Sunspot Watch, TE** Students study sunspots in their classroom by taping a piece of cardboard with a hole punched in the center to a window. **(ADVANCED)**

__ **Discussion, Comprehension Check, TE** Students discuss what the motion of sunspots tells scientists about how the sun's gases rotate. **(BASIC)**

__ **Using the Figure, Sunspot Cycles, TE** Students use the graph illustrating recorded sunspots to answer questions about sunspot cycles. **(GENERAL)**

__ **Reading Skill Builder, Paired Summarizing, TE** Students read the text in pairs and take turns summarizing what they have read. **(BASIC)**

KEY **SE** = Student Edition **TE** = Teacher Edition **CRF** = Chapter Resource File

__ **Internet Activity, Solar Activity and Climate, TE** Students research the relationship between solar activity levels and climate on Earth. **(ADVANCED)**

__ **Internet Activity, Solar Activity and Climate, CRF** This worksheet directs student research for the topic presented in the Teacher Edition. **(ADVANCED)**

__ **Skill Builder, Writing, TE** Students write a passage about the light displays produced by auroras. **(GENERAL)**

__ **Group Activity, Geomagnetic Storms, TE** In groups, students explore methods of dealing with the effects of geomagnetic storms. **(GENERAL)**

__ **Impact on Society, The Genesis Mission, SE** This article discusses the Genesis Mission and scientists' plans to use the collected data to study solar wind.

__ **Discussion, Collecting Samples, TE** Students review the objectives of the *Genesis* project and compare them to the procedures used on the *Apollo* missions. **(BASIC)**

__ **Interactive Tutor CD-ROM, Hail, Northern Lights, and Rainbows** Students practice what they have learned using this interactive activity. **(GENERAL)**

__ **Interactive Tutor CD-ROM, Our Sun and Origin of the Solar System** Students practice what they have learned using this interactive activity. **(GENERAL)**

CLOSE *(10 minutes)*

__ **Section Review, SE** Students answer end-of-section vocabulary, key ideas, critical thinking, and concept mapping questions. **(GENERAL)**

__ **Reteaching, Solar Lingo, TE** In groups, students create study cards of the key terms in this section. **(BASIC)**

__ **Quiz, TE** Students answer 2 questions about this section. **(GENERAL)**

__ **Section Quiz, CRF** Use this quiz to assess students' understanding of the section. **(GENERAL) Also in Spanish**

__ **Spanish Section Quiz, Spanish Assessments** Use this quiz to assess students' understanding of the section. **(GENERAL)**

__ **Alternative Assessment, The Active Sun, TE** Students create posters explaining how electromagnetic radiation and particles from the sun affect life on Earth. **(ADVANCED)**

KEY **SE** = Student Edition **TE** = Teacher Edition **CRF** = Chapter Resource File

End of Chapter Review and Assessment

PACING

Regular Schedule: **with lab(s): N/A** **without lab(s): 2 days**
Block Schedule: **with lab(s): N/A** **without lab(s): 1 day**

__ **Chapter Review, SE** Students answer end-of-chapter vocabulary, key ideas, critical thinking, math, writing, and interpreting graphics questions. **(GENERAL)**

__ **Alternative Assessment, Our Daytime Star, TE** Students create models, demonstrations, and activities about the sun and to present to younger children. **(GENERAL)**

__ **Standardized Test Prep, SE** Have students complete the standardized test prep to help them prepare for standardized tests. **(GENERAL)**

__ **Holt PuzzlePro®** Use this software to create crossword puzzles and word searches that make learning vocabulary fun. **(GENERAL)**

__ **Concept Review, CRF** Use this worksheet to assess student's understanding of the main concepts presented in the chapter. **(GENERAL)**

__ **Critical Thinking, CRF** Students use this worksheet to develop and practice critical thinking skills. **(ADVANCED)**

__ **Math Skills, CRF** This worksheet helps reinforce students' understanding of scientific notation. **(GENERAL)**

__ **Graphing Skills, CRF** This worksheet helps reinforce students' understanding of bar graphs. **(GENERAL)**

__ **Concept Review, Study Guide, The Sun** Use this worksheet to assess student's understanding of the main concepts presented in the chapter. **(GENERAL)**

__ **Concept Review, Spanish Study Guide, The Sun** This worksheet helps reinforce English-language learners' understanding of the main concepts presented in the chapter. **(GENERAL)**

__ **Chapter Test A, CRF** Assign questions from the appropriate test for chapter assessment. **(GENERAL)**

__ **Chapter Test B, CRF** Assign questions from the appropriate test for chapter assessment. **(ADVANCED)**

__ **Spanish Chapter Test A, Spanish Assessments** Assign questions from the appropriate test for chapter assessment. **(GENERAL)**

__ **Test Generator, One-Stop Planner** Create a customized homework assignment, quiz, or test using the HRW Test Generator program. **(GENERAL)**

__ **Test Item Listing, CRF** Use the Test Item Listing to identify questions for a customized homework, quiz, or test.

__ **PowerPoint® Resources** Use the customizable presentation to help students practice standardized test-taking skills.

KEY **SE** = Student Edition **TE** = Teacher Edition **CRF** = Chapter Resource File

OTHER RESOURCE OPTIONS

__ **Brain Food Video Quiz** Use this video resource to assess student's progress. **(GENERAL)**

KEY **SE** = Student Edition **TE** = Teacher Edition **CRF** = Chapter Resource File

Chapter 30 Stars, Galaxies, and the Universe

Chapter Opener

__ **Pre-Reading Activity, Layered Book, SE**

__ **Using the Figure, Tarantula Nebula, TE (GENERAL)**

Section 1 Characteristics of Stars

PACING

Regular Schedule:	**with lab(s): 2 days**	**without lab(s): 1 day**
Block Schedule:	**with lab(s): 1 day**	**without lab(s): 0.5 day**

OBJECTIVES

1. **Describe** how astronomers determine the composition and temperature of stars.
2. **Explain** why stars appear to move in the sky.
3. **Describe** one way astronomers measure the distances to stars.
4. **Explain** the difference between absolute magnitude and apparent magnitude.

NATIONAL SCIENCE EDUCATION STANDARDS

SAI 2e: Scientific explanations must adhere to criteria such as: a proposed explanation must be logically consistent; it must abide by the rules of evidence; it must be open to questions and possible modification; and it must be based on historical and current scientific knowledge.

FOCUS *(5 minutes)*

__ **Overview, TE** Review the objectives listed in the Student Edition. **(GENERAL)**

__ **PowerPoint® Resources** Use the customizable presentation to enhance your lecture.

__ **Bellringer, TE** Students write down the names of stars that they are familiar with. **(GENERAL)**

__ **Bellringer Transparency** Use this transparency as students enter the classroom and find their seats.

MOTIVATE *(10 minutes)*

__ **Demonstration, Spectra, TE** Students observe a demonstration of the emission lines produced by hot gases. **(GENERAL)**

TEACH *(65 minutes)*

__ **Directed Reading, CRF** These worksheets reinforce basic concepts and vocabulary presented in the section. **(BASIC)**

KEY **SE** = Student Edition **TE** = Teacher Edition **CRF** = Chapter Resource File

__ **Internet Activity, Proper Motion, TE** Students research proper motion in stars. **(ADVANCED)**

__ **Internet Activity, Proper Motion, CRF** This worksheet directs student research for the topic presented in the Teacher Edition. **(ADVANCED)**

__ **Activity, Observing Stars, TE** Students use sketches to better understand which stars appear at different latitudes. **(ADVANCED)**

__ **Teaching Transparency 149, The Doppler Effect with Worksheet** Use this Teaching Transparency and the accompanying worksheet to support your classroom discussion of the Doppler Effect.

__ **Quick Lab, Parallax, SE** Students use paper plates and thread to examine the positions of stars and how they move. **(GENERAL)**

__ **Datasheet for Quick Lab, Parallax, CRF** Students use the datasheet to complete the Quick Lab. **(GENERAL)**

__ **Teaching Transparency 150, Apparent Magnitude with Worksheet** Use this Teaching Transparency and the accompanying worksheet to support your classroom discussion of apparent magnitude.

__ **Making Models Lab, Star Magnitudes, SE** Students use flashlights to see the effect of distance on brightness and the relationship between temperature and color. **(GENERAL)**

__ **Datasheet for Chapter Lab, Making Models Lab, Star Magnitudes, CRF** Students use the datasheet to complete the Chapter Lab. **(GENERAL)**

__ **Activity, Space Telescopes, TE** Students research space telescopes and propose projects for a space telescope based on an area that interests them. **(GENERAL)**

__ **Inquiry Lab, Curving Space-Time, CRF** Students examine how changes in mass affect the curvature of space-time. **(GENERAL)**

__ **Skills Practice Lab, Blackbody Radiation, CRF** Students relate the blackbody behavior of light bulbs to the temperature and brightness of stars. **(GENERAL)**

__ **Interactive Tutor CD-ROM, Distances** Students practice what they have learned using this interactive activity. **(GENERAL)**

CLOSE *(10 minutes)*

__ **Section Review, SE** Students answer end-of-section vocabulary, key ideas, critical thinking, and concept mapping questions. **(GENERAL)**

__ **Reteaching, Star Temperatures and Colors, TE** Students use colored cards to arrange stars by temperature from coolest to hottest. **(BASIC)**

__ **Quiz, TE** Students answer 3 questions about this section. **(GENERAL)**

__ **Section Quiz, CRF** Use this quiz to assess students' understanding of the section. **(GENERAL) Also in Spanish**

__ **Spanish Section Quiz, Spanish Assessments** Use this quiz to assess students' understanding of the section. **(GENERAL)**

__ **Alternative Assessment, Moving Stars, TE** Students create posters showing how a chosen constellation will move over time. **(ADVANCED)**

KEY **SE** = Student Edition **TE** = Teacher Edition **CRF** = Chapter Resource File

Section 2 Stellar Evolution

PACING

Regular Schedule: **with lab(s): N/A** **without lab(s): 2 days**
Block Schedule: **with lab(s): N/A** **without lab(s): 1 day**

OBJECTIVES

1. **Describe** how a protostar becomes a star.
2. **Explain** how a main-sequence star generates energy.
3. **Describe** the evolution of a star after its main-sequence stage.

NATIONAL SCIENCE EDUCATION STANDARDS

ES 4b: Early in the history of the universe, matter, primarily the light atoms hydrogen and helium, clumped together by gravitational attraction to form countless trillions of stars. Billions of galaxies, each of which is a gravitationally bound cluster of billions of stars, now form most of the visible mass in the universe.

ES 4c: Stars produce energy from nuclear reactions, primarily the fusion of hydrogen to form helium. These and other processes in stars have led to the formation of all the other elements.

SAI 2e: Scientific explanations must adhere to criteria such as: a proposed explanation must be logically consistent; it must abide by the rules of evidence; it must be open to questions and possible modification; and it must be based on historical and current scientific knowledge.

UCP 3: Although most things are in the process of becoming different—changing—some properties of objects and processes are characterized by constancy, including the speed of light, the charge of an electron, and the total mass plus energy in the universe. Changes might occur, for example, in properties of materials, position of objects, motion, and form and function of systems. Interactions within and among systems result in change. Changes vary in rate, scale, and pattern, including trends and cycles.

FOCUS *(5 minutes)*

__ **Overview, TE** Review the objectives listed in the Student Edition. **(GENERAL)**

__ **PowerPoint® Resources** Use the customizable presentation to enhance your lecture.

__ **Bellringer, TE** Students write down a few characteristics of stars that they think will change as a star ages. **(GENERAL)**

__ **Bellringer Transparency** Use this transparency as students enter the classroom and find their seats.

KEY **SE** = Student Edition **TE** = Teacher Edition **CRF** = Chapter Resource File

MOTIVATE *(10 minutes)*

__ **Using the Figure, Star Mass and Classification, TE** Students study the H-R diagram and note which properties vary along the horizontal and vertical axes. **(GENERAL)**

TEACH *(20 minutes)*

__ **Directed Reading, CRF** These worksheets reinforce basic concepts and vocabulary presented in the section. **(BASIC)**

__ **Teaching Transparency 151, The Hertzsprung-Russell Diagram with Worksheet** Use this Teaching Transparency and the accompanying worksheet to support your classroom discussion of the Hertzsprung-Russell diagram.

__ **Using the Figure, Masses of Stars, TE** Students discuss how the mass of material in a protostar determines what class of star will form. **(GENERAL)**

__ **Group Activity, Main-Sequence Stars, TE** In groups, students research the properties of each class of main-sequence star. **(ADVANCED)**

__ **Using the Figure, Identifying Stars, TE** Students study the image of the constellation Orion and discuss why the stars are so bright. **(BASIC)**

__ **Physics Connection, Star Size and Spectra, TE** Students research how spectra are used to identify the size and temperature of stars. **(GENERAL)**

__ **Reading Skill Builder, Paired Summarizing, TE** Students read the text in pairs and take turns summarizing what they have read. **(BASIC)**

__ **Debate, SETI, TE** Students research and debate the pros and cons of the SETI program. **(ADVANCED)**

__ **Graphic Organizer, Chain-of-Events Chart, SE** Use the graphic organizer to help students develop their analytical skills. **(GENERAL)**

__ **Using the Figure, Identifying Stars, TE** Students discuss what causes the differences between the upper and lower paths of stars. **(BASIC)**

__ **Teaching Transparency 152, The Life Cycle of Stars with Worksheet** Use this Teaching Transparency and the accompanying worksheet to support your classroom discussion of the life cycle of stars.

__ **Skill Builder, Vocabulary, TE** Students discuss the Latin meanings of the words *nova, supernova* and *nebula.* **(GENERAL)**

__ **Physics Connection, Supernova Physics, TE** Students research the types and properties of supernovas. **(ADVANCED)**

__ **Interactive Tutor CD-ROM, Stars** Students practice what they have learned using this interactive activity. **(GENERAL)**

CLOSE *(10 minutes)*

__ **Section Review, SE** Students answer end-of-section vocabulary, key ideas, critical thinking, and concept mapping questions. **(GENERAL)**

KEY **SE** = Student Edition **TE** = Teacher Edition **CRF** = Chapter Resource File

__ **Homework, Star Remnants, TE** Students research the discoveries of the first white dwarf and neutron stars. **(GENERAL)**

__ **Reteaching, Stellar Evolution, TE** Students categorize the phases of a star's life in time order. **(BASIC)**

__ **Quiz, TE** Students answer 2 questions about this section. **(GENERAL)**

__ **Section Quiz, CRF** Use this quiz to assess students' understanding of the section. **(GENERAL) Also in Spanish**

__ **Spanish Section Quiz, Spanish Assessments** Use this quiz to assess students' understanding of the section. **(GENERAL)**

__ **Alternative Assessment, Odd Stars, TE** Students research the evolution of Wolf-Rayet stars and blue stragglers. **(ADVANCED)**

KEY **SE** = Student Edition **TE** = Teacher Edition **CRF** = Chapter Resource File

Section 3 Star Groups

PACING

Regular Schedule:	**with lab(s): N/A**	**without lab(s): 2 days**
Block Schedule:	**with lab(s): N/A**	**without lab(s): 1 day**

OBJECTIVES

1. **Describe** the characteristics that identify a constellation.
2. **Describe** the three main types of galaxies.
3. **Explain** how a quasar differs from a typical galaxy.

NATIONAL SCIENCE EDUCATION STANDARDS

SAI 2e: Scientific explanations must adhere to criteria such as: a proposed explanation must be logically consistent; it must abide by the rules of evidence; it must be open to questions and possible modification; and it must be based on historical and current scientific knowledge.

UCP 1: The natural and designed world is complex; it is too large and complicated to investigate and comprehend all at once. Scientists and students learn to define small portions for the convenience of investigation. The units of investigation can be referred to as "systems." A system is an organized group of related objects or components that form a whole. Systems can consist, for example, of organisms, machines, fundamental particles, galaxies, ideas, numbers, transportation, and education. Systems have boundaries, components, resources flow (input and output), and feedback.

FOCUS *(5 minutes)*

__ **Overview, TE** Review the objectives listed in the Student Edition. **(GENERAL)**

__ **PowerPoint® Resources** Use the customizable presentation to enhance your lecture.

__ **Bellringer, TE** Students write down the names of five constellations that they know. **(GENERAL)**

__ **Bellringer Transparency** Use this transparency as students enter the classroom and find their seats.

MOTIVATE *(10 minutes)*

__ **Activity, Constellations, TE** Students research a constellation's name, history, and cultural significance (other than Greek or Roman). **(GENERAL)**

TEACH *(20 minutes)*

__ **Directed Reading, CRF** These worksheets reinforce basic concepts and vocabulary presented in the section. **(BASIC)**

KEY **SE** = Student Edition **TE** = Teacher Edition **CRF** = Chapter Resource File

__ **Teaching Transparency 153, The Constellation Orion with Worksheet** Use this Teaching Transparency and the accompanying worksheet to support your classroom discussion of the constellation Orion.

__ **Skill Builder, Writing, TE** Students research stars that create open clusters and globular clusters. **(ADVANCED)**

__ **Astronomy Connection, Identifying a Galaxy from Within, TE** Students research methods that have been used to map the space around the solar system. **(ADVANCED)**

__ **Reading Skill Builder, Anticipation Guide, TE** Students predict differences between spiral, barred spiral, elliptical, and irregular galaxies before reading the page. **(BASIC)**

__ **Activity, Multimedia Project, TE** Students research the four basic classes of galaxies. **(GENERAL)**

__ **Maps in Action, The Milky Way, SE** Students enhance their map-reading skills by answering questions about a map showing the Milky Way. **(GENERAL)**

__ **Teaching Transparency 155, The Milky Way with Worksheet** Use this Teaching Transparency and the accompanying worksheet to help students perform the Maps in Action activity.

__ **Mapping Expeditions, Stars in Your Eyes, SE** Students examine a map showing the constellations visible from the northern hemisphere. **(GENERAL)**

__ **Group Activity, Charting the Galaxy, TE** In groups, students research portions of the Milky Way's arms and create maps of their sections. **(GENERAL)**

__ **Science and Technology, Studying Stars in Formation, SE** The contributions of the Hubble and the Spitzer Telescopes to the study of stars and galaxies are discussed.

CLOSE *(10 minutes)*

__ **Section Review, SE** Students answer end-of-section vocabulary, key ideas, critical thinking, and concept mapping questions. **(GENERAL)**

__ **Reteaching, Galaxy Types, TE** Students use flashcards to identify which galaxy type the photos represent. **(BASIC)**

__ **Quiz, TE** Students answer 3 questions about this section. **(GENERAL)**

__ **Section Quiz, CRF** Use this quiz to assess students' understanding of the section. **(GENERAL) Also in Spanish**

__ **Spanish Section Quiz, Spanish Assessments** Use this quiz to assess students' understanding of the section. **(GENERAL)**

__ **Alternative Assessment, Modeling Constellations, TE** Students create models that show the actual relative positions of stars in a constellation. **(ADVANCED)**

KEY **SE** = Student Edition **TE** = Teacher Edition **CRF** = Chapter Resource File

Section 4 The Big Bang Theory

PACING

Regular Schedule: **with lab(s): N/A** **without lab(s): 2 days**
Block Schedule: **with lab(s): N/A** **without lab(s): 1 day**

OBJECTIVES

1. **Explain** how Hubble's discoveries lead to an understanding that the universe is expanding.
2. **Summarize** the big bang theory.
3. **List** evidence for the big bang theory.

NATIONAL SCIENCE EDUCATION STANDARDS

ES 4a: The origin of the universe remains one of the greatest questions in science. The "big bang" theory places the origin between 10 and 20 billion years ago, when the universe began in a hot dense state; according to this theory, the universe has been expanding ever since.

SAI 2e: Scientific explanations must adhere to criteria such as: a proposed explanation must be logically consistent; it must abide by the rules of evidence; it must be open to questions and possible modification; and it must be based on historical and current scientific knowledge.

FOCUS *(5 minutes)*

__ **Overview, TE** Review the objectives listed in the Student Edition. **(GENERAL)**

__ **PowerPoint® Resources** Use the customizable presentation to enhance your lecture.

__ **Bellringer, TE** Students discuss how they think the universe was formed. **(GENERAL)**

__ **Bellringer Transparency** Use this transparency as students enter the classroom and find their seats.

MOTIVATE *(10 minutes)*

__ **Discussion, The Expanding Universe, TE** Students recall how the Doppler effect relates to light from a moving object. **(GENERAL)**

TEACH *(20 minutes)*

__ **Directed Reading, CRF** These worksheets reinforce basic concepts and vocabulary presented in the section. **(BASIC)**

__ **Teaching Transparency 154, Timeline of the Big Bang with Worksheet** Use this Teaching Transparency and the accompanying worksheet to support your classroom discussion of the timeline of the big bang.

KEY **SE** = Student Edition **TE** = Teacher Edition **CRF** = Chapter Resource File

__ **Physics Connection, New Physics, Old Universe, TE** Students research scientific models of the universe leading up to the current model. **(ADVANCED)**

__ **Quick Lab, The Expanding Universe, SE** Students use balloons to illustrate how galaxies move. **(GENERAL)**

__ **Datasheet for Quick Lab, The Expanding Universe, CRF** Students use the datasheet to complete the Quick Lab. **(GENERAL)**

__ **Interactive Tutor CD-ROM, Galaxies and the Universe** Students practice what they have learned using this interactive activity. **(GENERAL)**

CLOSE *(10 minutes)*

__ **Section Review, SE** Students answer end-of-section vocabulary, key ideas, critical thinking, and concept mapping questions. **(GENERAL)**

__ **Reteaching, Timeline, TE** Students create a timeline that shows the major events in the development of the universe according to the big bang theory. **(BASIC)**

__ **Quiz, TE** Students answer 3 questions about this section. **(GENERAL)**

__ **Section Quiz, CRF** Use this quiz to assess students' understanding of the section. **(GENERAL) Also in Spanish**

__ **Spanish Section Quiz, Spanish Assessments** Use this quiz to assess students' understanding of the section. **(GENERAL)**

__ **Alternative Assessment, Rate of Universal Expansion, TE** Students research the Hubble constant. **(ADVANCED)**

KEY **SE** = Student Edition **TE** = Teacher Edition **CRF** = Chapter Resource File

End of Chapter Review and Assessment

PACING

Regular Schedule: **with lab(s): N/A** **without lab(s): 2 days**
Block Schedule: **with lab(s): N/A** **without lab(s): 1 day**

__ **Chapter Review, SE** Students answer end-of-chapter vocabulary, key ideas, critical thinking, math, writing, and interpreting graphics questions. **(GENERAL)**

__ **Alternative Assessment, Poster Project, TE** Students make posters showing classes of stars, their properties, and regions where other types of stars might be found. **(GENERAL)**

__ **Standardized Test Prep, SE** Have students complete the standardized test prep to help them prepare for standardized tests. **(GENERAL)**

__ **Holt PuzzlePro®** Use this software to create crossword puzzles and word searches that make learning vocabulary fun. **(GENERAL)**

__ **Concept Review, CRF** Use this worksheet to assess student's understanding of the main concepts presented in the chapter. **(GENERAL)**

__ **Critical Thinking, CRF** Students use this worksheet to develop and practice their critical thinking skills. **(ADVANCED)**

__ **Math Skills, CRF** This worksheet helps reinforce students' understanding of order of operations. **(GENERAL)**

__ **Graphing Skills, CRF** This worksheet helps reinforce students' understanding of line graphs. **(GENERAL)**

__ **Concept Review, Study Guide, Stars, Galaxies, and the Universe** Use this worksheet to assess student's understanding of the main concepts presented in the chapter. **(GENERAL)**

__ **Concept Review, Spanish Study Guide, Stars, Galaxies, and the Universe** This worksheet helps reinforce English-language learners' understanding of the main concepts presented in the chapter. **(GENERAL)**

__ **Chapter Test A, CRF** Assign questions from the appropriate test for chapter assessment. **(GENERAL)**

__ **Chapter Test B, CRF** Assign questions from the appropriate test for chapter assessment. **(ADVANCED)**

__ **Spanish Chapter Test A, Spanish Assessments** Assign questions from the appropriate test for chapter assessment. **(GENERAL)**

__ **Test Generator, One-Stop Planner** Create a customized homework assignment, quiz, or test using the HRW Test Generator program. **(GENERAL)**

__ **Test Item Listing, CRF** Use the Test Item Listing to identify questions for a customized homework, quiz, or test.

__ **PowerPoint® Resources** Use the customizable presentation to help students practice standardized test-taking skills.

KEY **SE** = Student Edition **TE** = Teacher Edition **CRF** = Chapter Resource File

OTHER RESOURCE OPTIONS

__ **NOVA Video, "Runaway Universe"** Join two competing teams in their quest to discover the secrets to the stars and the ultimate fate of the universe.

__ **Brain Food Video Quiz** Use this video resource to assess student's progress. **(GENERAL)**

KEY SE = Student Edition **TE** = Teacher Edition **CRF** = Chapter Resource File